Methods in Molecular Biology

Series Editor
John M. Walker
School of Life and Medical Sciences
University of Hertfordshire
Hatfield, Hertfordshire, UK

For further volumes:
http://www.springer.com/series/7651

For over 35 years, biological scientists have come to rely on the research protocols and methodologies in the critically acclaimed *Methods in Molecular Biology* series. The series was the first to introduce the step-by-step protocols approach that has become the standard in all biomedical protocol publishing. Each protocol is provided in readily-reproducible step-by step fashion, opening with an introductory overview, a list of the materials and reagents needed to complete the experiment, and followed by a detailed procedure that is supported with a helpful notes section offering tips and tricks of the trade as well as troubleshooting advice. These hallmark features were introduced by series editor Dr. John Walker and constitute the key ingredient in each and every volume of the *Methods in Molecular Biology* series. Tested and trusted, comprehensive and reliable, all protocols from the series are indexed in PubMed.

RNA Structure and Dynamics

Edited by

Jienyu Ding, Jason R. Stagno, and Yun-Xing Wang

Protein-Nucleic Acid Interaction Section, Center for Structural Biology, Center for Cancer Research, National Cancer Institute, Frederick, MD, USA

Humana Press

Editors
Jienyu Ding
Protein-Nucleic Acid Interaction
Section, Center for Structural Biology,
Center for Cancer Research
National Cancer Institute
Frederick, MD, USA

Jason R. Stagno
Protein-Nucleic Acid Interaction
Section, Center for Structural Biology,
Center for Cancer Research
National Cancer Institute
Frederick, MD, USA

Yun-Xing Wang
Protein-Nucleic Acid Interaction
Section, Center for Structural Biology,
Center for Cancer Research
National Cancer Institute
Frederick, MD, USA

ISSN 1064-3745 ISSN 1940-6029 (electronic)
Methods in Molecular Biology
ISBN 978-1-0716-2689-4 ISBN 978-1-0716-2687-0 (eBook)
https://doi.org/10.1007/978-1-0716-2687-0

This Humana imprint is published by the registered company Springer Science+Business Media, LLC, part of Springer Nature.
The registered company address is: 1 New York Plaza, New York, NY 10004, U.S.A.

Preface

From its infancy, when the phrase "The RNA World" was first mentioned in 1962 and the first tRNA structure was determined in 1974, both by Alex Rich, to the pinnacle, first complete structure of the large 50S ribosomal subunit by Tom Steitz in 2000, our understanding of RNA structural biology has led us to realize that RNA structure and dynamics are really the two sides of the same coin, both of which are relevant to RNA function. The scope of the chapters in this book is to describe and illustrate various approaches and methods together with detailed experimental and computational procedures, all of which are essential in actual projects studying RNA structure and dynamics. This book compiles the best collection of the near-complete and comprehensive protocols for studies of RNA structure and dynamics using a variety of technologies, from the latest technologies, such as X-ray free electron lasers and cryo-electron microscopy, to main staple tools or methods, such as crystallography, isothermal titration calorimetry, small angle X-ray scattering, single-molecule Förster Energy transfer, and atomic force microscopy, as well as more sophisticated computational simulation and prediction. All contributing authors are either leaders or emerging stars in these important areas of research. In particular, I thank all authors for their contributions to this timely book under the very challenging circumstance when almost every aspect of daily work and life was interrupted during the COVID-19 pandemic. Without their dedication to basic science research, this book would not have been possible. Finally, I would like to thank my co-editors, Drs. Jienyu Ding and Jason Stagno, for their assistance in checking every detail.

Frederick, MD, USA *Yun-Xing Wang*

Contents

Contributors

Ascensión Ariza-Mateos • *Department of Biochemistry and Molecular Pharmacology, New York University Grossman School of Medicine, New York, NY, USA*

Lei Bao • *School of Public Health, Hubei University of Medicine, Shiyan, China*

Chapin E. Cavender • *Department of Biochemistry & Biophysics, University of Rochester School of Medicine & Dentistry, Rochester, NY, USA; Center for RNA Biology, University of Rochester School of Medicine & Dentistry, Rochester, NY, USA*

Henry N. Chapman • *Center for Free-Electron Laser Science, Deutsches Elektronen-Synchrotron DESY, Hamburg, Germany; Department of Physics, Universität Hamburg, Hamburg, Germany; Centre for Ultrafast Imaging, Universität Hamburg, Hamburg, Germany*

Binxian Chen • *Beijing Advanced Innovation Center for Structural Biology, School of Life Sciences, Tsinghua University, Beijing, China*

Wah Chiu • *Department of Bioengineering, James H. Clark Center, Stanford University, Stanford, CA, USA; CryoEM and Bioimaging Division, Stanford Synchrotron Radiation Lightsource, SLAC National Accelerator Laboratory, Stanford University, Menlo Park, CA, USA*

Rhiju Das • *Biophysics Program, Stanford University, Stanford, CA, USA*

Jienyu Ding • *Protein-Nucleic Acid Interaction Section, Center for Structural Biology, Center for Cancer Research, National Cancer Institute, Frederick, MD, USA*

Lixin Fan • *Basic Science Program, Frederick National Laboratory for Cancer Research, National Cancer Institute, Frederick, MD, USA*

Xianyang Fang • *Beijing Advanced Innovation Center for Structural Biology, School of Life Sciences, Tsinghua University, Beijing, China*

Adrian R. Ferré-D'Amaré • *Biochemistry and Biophysics Center, National Heart, Lung and Blood Institute, Bethesda, MD, USA*

Wenze Huang • *MOE Key Laboratory of Bioinformatics, Center for Synthetic and Systems Biology, School of Life Sciences, Tsinghua University, Beijing, China; Beijing Advanced Innovation Center for Structural Biology & Frontier Research Center for Biological Structure, School of Life Sciences, Tsinghua University, Beijing, China; Tsinghua-Peking Center for Life Sciences, Beijing, China*

Kalli Kappel • *Biophysics Program, Stanford University, Stanford, CA, USA*

Wojciech K. Kasprzak • *Basic Research Laboratory, Frederick National Laboratory for Cancer Research, National Cancer Institute, Frederick, MD, USA*

Juraj Knoska • *Center for Free-Electron Laser Science, Deutsches Elektronen-Synchrotron DESY, Hamburg, Germany; Department of Physics, Universität Hamburg, Hamburg, Germany*

Jay R. Knutson • *Biochemistry and Biophysics Center, National Heart, Lung and Blood Institute, Bethesda, MD, USA*

Yun-Tzai Lee • *Protein-Nucleic Acid Interaction Section, Center for Structural Biology, Center for Cancer Research, National Cancer Institute, Frederick, MD, USA*

Shanshan Li • *MOE Key Laboratory for Cellular Dynamics and Division of Life Sciences and Medicine, University of Science and Technology of China, Hefei, China; Department of Bioengineering, James H. Clark Center, Stanford University, Stanford, CA, USA*

SHUANG LI • *Laboratory of Molecular Biology, National Institute of Diabetes and Digestive and Kidney Diseases, Bethesda, MD, USA; Structural Biochemistry Unit, National Institute of Dental and Craniofacial Research, Bethesda, MD, USA*
KATIE A. LINK • *Biochemistry and Biophysics Center, National Heart, Lung and Blood Institute, Bethesda, MD, USA*
BINGNAN LUO • *The State Key Laboratory of Biotherapy, Department of Geriatrics and National Clinical Research Center for Geriatrics, West China Hospital, Sichuan University, Chengdu, Sichuan, China*
HAIYUN MA • *The State Key Laboratory of Biotherapy, Department of Geriatrics and National Clinical Research Center for Geriatrics, West China Hospital, Sichuan University, Chengdu, Sichuan, China*
DAVID H. MATHEWS • *Department of Biochemistry & Biophysics, University of Rochester School of Medicine & Dentistry, Rochester, NY, USA; Center for RNA Biology, University of Rochester School of Medicine & Dentistry, Rochester, NY, USA*
ASHOK NUTHANAKANTI • *Department of Biochemistry and Molecular Pharmacology, New York University Grossman School of Medicine, New York, NY, USA*
PHILLIP PHAM • *Biophysics Program, Stanford University, Stanford, CA, USA*
RAMYA RANGAN • *Biophysics Program, Stanford University, Stanford, CA, USA*
KRISHNA P. SAPKOTA • *Laboratory of Molecular Biology, National Institute of Diabetes and Digestive and Kidney Diseases, Bethesda, MD, USA*
ANDREAS SCHMIDT • *Department of Chemistry, Single Molecule Analysis Group and Center for RNA Biomedicine, University of Michigan, Ann Arbor, MI, USA*
GRIFFIN M. SCHROEDER • *Department of Biochemistry & Biophysics, University of Rochester School of Medicine & Dentistry, Rochester, NY, USA; Center for RNA Biology, University of Rochester School of Medicine & Dentistry, Rochester, NY, USA*
ALEXANDER SERGANOV • *Department of Biochemistry and Molecular Pharmacology, New York University Grossman School of Medicine, New York, NY, USA*
BRUCE A. SHAPIRO • *RNA Biology Laboratory, Center for Cancer Research, National Cancer Institute, Frederick, MD, USA*
JASON R. STAGNO • *Protein-Nucleic Acid Interaction Section, Center for Structural Biology, Center for Cancer Research, National Cancer Institute, Frederick, MD, USA*
ZHAOMING SU • *The State Key Laboratory of Biotherapy, Department of Geriatrics and National Clinical Research Center for Geriatrics, West China Hospital, Sichuan University, Chengdu, Sichuan, China*
ROBERT J. TRACHMAN III • *Biochemistry and Biophysics Center, National Heart, Lung and Blood Institute, Bethesda, MD, USA*
NILS G. WALTER • *Department of Chemistry, Single Molecule Analysis Group and Center for RNA Biomedicine, University of Michigan, Ann Arbor, MI, USA*
YUN-XING WANG • *Protein-Nucleic Acid Interaction Section, Center for Structural Biology, Center for Cancer Research, National Cancer Institute, Frederick, MD, USA*
JOSEPH E. WEDEKIND • *Department of Biochemistry & Biophysics, University of Rochester School of Medicine & Dentistry, Rochester, NY, USA; Center for RNA Biology, University of Rochester School of Medicine & Dentistry, Rochester, NY, USA*
ROBB WELTY • *Department of Chemistry, Single Molecule Analysis Group and Center for RNA Biomedicine, University of Michigan, Ann Arbor, MI, USA*
YI XIAO • *Institute of Biophysics, School of Physics, Huazhong University of Science and Technology, Wuhan, China*

Jie Zhang • *Beijing Advanced Innovation Center for Structural Biology, School of Life Sciences, Tsinghua University, Beijing, China*

Jinwei Zhang • *Laboratory of Molecular Biology, National Institute of Diabetes and Digestive and Kidney Diseases, Bethesda, MD, USA*

Kaiming Zhang • *MOE Key Laboratory for Cellular Dynamics and Division of Life Sciences and Medicine, University of Science and Technology of China, Hefei, China; Department of Bioengineering, James H. Clark Center, Stanford University, Stanford, CA, USA*

Qiangfeng Cliff Zhang • *MOE Key Laboratory of Bioinformatics, Center for Synthetic and Systems Biology, School of Life Sciences, Tsinghua University, Beijing, China; Beijing Advanced Innovation Center for Structural Biology & Frontier Research Center for Biological Structure, School of Life Sciences, Tsinghua University, Beijing, China; Tsinghua-Peking Center for Life Sciences, Beijing, China*

Chapter 1

Cotranscriptional Assembly and Native Purification of Large RNA–RNA Complexes for Structural Analyses

Krishna P. Sapkota, Shuang Li, and Jinwei Zhang

Abstract

Recent technological developments such as cryogenic electron microscopy (Cryo-EM) and X-ray free electron lasers (XFEL) have significantly expanded the available toolkit to visualize large, complex noncoding RNAs and their complexes. Consequently, the quality of the RNA sample, as measured by its chemical monodispersity and conformational homogeneity, has become the bottleneck that frequently precludes effective structural analyses. Here we describe a general RNA sample preparation protocol that combines cotranscriptional RNA folding and RNA–RNA complex assembly, followed by native purification of stoichiometric complexes. We illustrate and discuss the utility of this versatile method in overcoming RNA misfolding and enabling the structural and mechanistic elucidations of the T-box riboswitch–tRNA complexes.

Key words Transcription, Cotranscriptional folding, RNA–RNA interactions, Riboswitches, tRNA, T-box

1 Introduction

Like their protein counterparts, many large, complex noncoding RNAs rely on their three-dimensional structures to execute a wide range of cellular functions such as gene regulation (*e.g.*, riboswitches [1, 2]), RNA processing (*e.g.*, ribozymes [3–5]), and protein synthesis (*e.g.*, ribosomes [6, 7]). The inherently dynamic, programmable conformations that RNA can readily and predictably access make them excellent platforms to host regulatory devices [8, 9]. However, the same conformational flexibility also creates structural heterogeneities that render them challenging targets for high-resolution analyses [10–12]. With the recent advent of direct

Krishna P. Sapkota and Shuang Li contributed equally to this work.

Jienyu Ding et al. (eds.), *RNA Structure and Dynamics*, Methods in Molecular Biology, vol. 2568,
https://doi.org/10.1007/978-1-0716-2687-0_1,

electron detectors for Cryo-EM and other technical advances in structural biology, the rate-limiting step for visualizing large noncoding RNAs has shifted to sample preparation.

Typically, RNA samples under structural analyses are synthesized in vitro by high-yield, noncognate RNA polymerases (RNAPs) such as the T7 RNAP. These enzymes are dramatically faster than the cognate multisubunit RNAPs that normally transcribe and vectorially fold the RNAs of interest in cells [13, 14], and generally ignore pause signals that can be essential for hierarchical folding of complex structured RNA, especially those that harbor long-range interactions [15]. The challenge of correctly folding complex RNAs in vitro is only exacerbated by the difficulty to detect misfolded RNAs after the fact. Unlike proteins, misfolded RNAs do not usually form aggregates due to their high hydrophilicity and extensive negative charges that discourage inter-particle interactions. As a result, it can be very difficult to detect RNA misfolding, unless an in vitro functional assay is readily available, such as for ribozymes like the RNase P [16]. Finally, refolding complex structured RNAs after initial denaturation is not always successful, akin to refolding proteins from inclusion bodies. For example, the full-length *B. subtilis glyQS* T-box riboswitch RNA is prone to misfolding, likely due to the absence of a cognate tRNA partner that is normally present during cotranscriptional folding and complex formation in cells. Interestingly, a number of long noncoding RNAs such as HOTAIR exhibit strong tendencies to misfold and form large heterogeneous aggregates in vitro [12].

To help overcome these significant challenges in RNA misfolding and heterogeneity, we illustrate a general protocol below that proved essential to the preparation of RNA samples of sufficient quality, which enabled structural analyses of T-box riboswitch–tRNA complexes using X-ray crystallography, Cryo-EM, and Small-angle X-ray Scattering (SAXS) [17–19]. The method starts with cotranscriptional folding and assembly of the mRNA–tRNA complex, where the misfolding-prone T-box RNA is transcribed in the presence of its cognate tRNA partner, thus facilitating the formation of functionally relevant complexes. Then the method diverges depending on the size and stability of the complexes. With the smaller (~46 kDa) and less stable T-box discriminator-tRNA complex ($K_d \sim 2\ \mu M$), size-exclusion chromatography (SEC) is effective in isolating the 1:1 complex without dissociating it. With the larger (~ 78 kDa) and highly stable full-length T-box-tRNA complex ($K_d < 100$ nM), strong anion exchange chromatography on a Mono Q column was key to remove misfolded T-box RNA as well as oligomeric RNAs to obtain highly homogeneous samples for Cryo-EM analysis.

2 Materials

1. DNA oligonucleotides for PCR amplification (Integrated DNA Technologies).
2. Diethyl pyrocarbonate (DEPC, Sigma) - treated water.
3. Taq DNA polymerase, 5 U/μL. (New England Biolabs).
4. Deoxynucleotide triphosphates mix containing dATP, dGTP, dCTP, and dTTP, 25 mM each (Thermo Scientific).
5. Programmable thermal cycler (Bio-Rad).
6. T7 RNA polymerase, 50 U/μL (New England Biolabs).
7. 10× T7 transcription buffer: 300 mM Tris pH 8.1, 250 mM $MgCl_2$, 100 mM DTT, 20 mM spermidine, and 0.01% Triton X-100.
8. ATP, GTP, GMP, CTP, and UTP, 100 mM solution each in DEPC water (Millipore-Sigma).
9. Inorganic Pyrophosphatase (Millipore-Sigma).
10. RNA binding buffer: 50 mM HEPES-KOH, pH 7.0, 100 mM KCl, 20 mM $MgCl_2$, 5 mM tris (2-carboxyethyl) phosphine (TCEP).
11. SYBR Gold Nucleic Acid Gel Stain (Life technologies).
12. Ethidium Bromide (EtBr) Dye (Millipore-Sigma).
13. Nanodrop 2000c spectrophotometer (Thermo Scientific).
14. Benchtop centrifuge: Eppendorf 5424 and 5430R with F-35-6-30 rotor.
15. Amicon 10 kDa MWCO centrifugal filters (15 ml and 0.5 ml, Millipore-Sigma).
16. Ultrafree-MC Centrifugal Filter Units (Millipore-Sigma).
17. Vertical gel electrophoresis apparatus (C.B.S. Scientific).
18. 40% acrylamide solution (acrylamide: bis-acrylamide 29:1, Thermo Scientific)
19. Ammonium persulfate (APS, Sigma).
20. PowerPac Hv power supply (Bio-rad).
21. Gel-imaging system such as an ImageQuant 800 (Cytiva) or a Typhoon scanner (Cytiva).
22. AKTA Pure 25 FPLC system with F9-C fraction collector (Cytiva).
23. Superdex 200 Increase 10/300 GL column (Cytiva).
24. MonoQ 5/50 GL column (Cytiva).
25. All other chemicals from Sigma-Aldrich unless otherwise specified.

3 Methods

3.1 Choice of T7 Promoter and Desired RNA 5′ Phosphorylation State

1. T7 RNA polymerase (RNAP) initiates transcription de novo on T7 promoters. Two classes of T7 promoters namely ϕ2.5 (class II) and ϕ6.5 (class III) are commonly used. The ϕ6.5 class-III promoter has a sequence 5′-TAATACGACGACATATAG, where the final G is the first nucleotide of the synthesized RNA transcript. The ϕ2.5 class-II promoter initiates transcription with ATP and has a typical sequence of 5-′-TAATACGACGACATATTA, where the final A is the transcription start site [20]. Both ϕ6.5 and ϕ2.5 class promoters yield higher amount of RNA when the initiator nucleotide (GTP for ϕ6.5 and ATP for ϕ2.5) is followed by one or two Gs. The ϕ6.5 promoter is used as the *B. subtilis* glyQS T-box riboswitch and its cognate ligand *B. subtilis* $tRNA^{Gly}$ both have a G at the +1 position.
2. RNA transcripts synthesized by in vitro transcription bear a 5′ triphosphate (PPP) group. The fact that T7 RNAP does not recognize structural features of the initiating nucleotide beyond the GMP/AMP moiety allows convenient 5′ labeling of RNA with fluorophores, biotin, or cofactors like NAD, FAD, and CoA using a GMP/AMP derivative [21]. To mimic natural tRNAs which bear a 5′-monophosphate, GMP priming is used to generate $tRNA^{Gly}$, where a GMP:GTP ratio of 10:1 is used, producing a ~ 90% 5′-monophosphorylated tRNA sample. If a 5′-OH group is desired on the transcript, a cis-cleaving Hammerhead ribozyme is appended 5′ to the RNA of interest to produce a nonphosphorylated RNA 5′-end (see the section below).
3. T7 RNAP occasionally produces 5′ heterogeneities and frequently generates 3′ heterogeneities in the form of N + 1, N + 2, or even longer transcripts [22]. In the next section, we describe a previously reported method that utilizes a cis-cleaving Hammerhead ribozyme and a trans-cleaving Varkud satellite (VS) ribozyme to overcome the 5′ and 3′ heterogeneities that can hamper structural analyses [23].

3.2 (Optional) Homogenization of RNA Termini by Dual Ribozyme Processing

1. The use of a cis-acting Hammerhead ribozyme and a trans-acting VS ribozyme allows facile synthesis of chemically homogenous RNA transcripts with defined 5′ and 3′ ends. These optional transcript-processing cassettes that flank the RNA of interest can be included in initial gene synthesis design so that they are available if needed. Cleavage by the 5-′-appended Hammerhead ribozyme leaves a 5′-OH group on the downstream target RNA, whereas cleavage and removal of the 3′-appended VS substrate RNA (~24 nucleotides, or nts)

by the trans-acting VS ribozyme leaves a 2′,3′-cyclic phosphate on the 3′ end of the target RNA. If needed, the cyclic phosphate can be converted to a linear 3′-phosphate by hydrochloric acid (HCl) treatment [18, 24]. Here, the VS ribozyme is provided either in the form of a restriction enzyme-linearized plasmid that encodes a T7 promoter-driven VS gene near its downstream open end, PCR products of a similar segment, or pre-transcribed VS RNA.

2. As an alternative to the VS ribozyme cleavage approach, RNA transcripts with homogenous 3′-ends can generally be produced using DNA templates that bear two consecutive 2′-O-methyl modifications at the 5′- terminus of the DNA template strand [9, 22, 25]. When PCR amplification is used to generate DNA templates, these 2′-O-methyl modifications can be provided by the reverse primer.

3.3 Streamlined DNA Template Preparation and Cotranscriptional Assembly of RNA–RNA Complexes

1. Assemble a 1 mL PCR reaction by mixing 100 μL 10× Taq buffer, forward and reverse primers to 5 μM final concentration, 100 μL 2.5 mM each dNTP, 5 μL Taq DNA Polymerase (DNAP, 25 units), and 1 μg plasmid DNA. 10× Taq buffer: 100 mM Tris-Cl pH 8.0, 500 mM KCl, 15 mM $MgCl_2$. Verify the expected length of PCR products by agarose gel electrophoresis. Testing showed that the PCR reaction can be used directly as a template for in vitro transcription in its entirety without further purification. Removal of the Taq DNAP by phenol/chloroform extraction or PCR product purification is unnecessary.
2. With appropriate DNA templates in hand, which can include several PCR products or linearized plasmids that encode the target RNA and any applicable trans-acting ribozymes, proceed to analytical in vitro transcription trials. When multiple DNA templates are included in the same reaction, they compete for the same pool of T7 RNAP and exhibit different efficiencies in T7 utilization and RNA yield. To balance the relative yield of the transcripts, such as the target RNA bearing the VS substrate appendage relative to the VS ribozyme, or the tRNA relative to the T-box RNA, several ratios of the DNA templates are generally tested to optimize the production of end-processed target RNA, or stoichiometric RNA–RNA complex formation (*see* **Note 1**).
3. Assemble several small-scale (50 μL) in vitro transcription reactions to optimize complex assembly. Scale up to 1–10 mL for preparative transcription reactions. 1 mg of purified RNA can be reasonably obtained from a 1 mL transcription reaction. For a standard 10 mL reaction, mix 1 mL 10× T7 transcription buffer, 25 μL 2 M $MgCl_2$ (to 5 mM), 500 μL each of 100 mM ATP, CTP, GTP, and UTP (to 5 mM), up to 4 mL (40% v/v) of

PCR reaction, an appropriate amount of T7 RNAP and inorganic pyrophosphatase, and use DEPC water to bring up to volume. The optimal quantities of the two enzymes are usually empirically determined by titration series that monitor RNA yield using Urea-PAGE. 10× T7 transcription buffer: 300 mM Tris-HCl pH 8.1, 250 mM $MgCl_2$, 100 mM dithiothreitol (DTT), 0.1% Triton X-100, and 20 mM spermidine in DEPC water. Incubate at 37 °C for 2–12 hours in a water bath or PCR machine.

4. Prepare a nondenaturing (or native) gel solution and cast the analytical gel while waiting for the transcription reaction. To make 40 mL 10% nondenaturing polyacrylamide gel solution, mix 8 mL 5× THE (Tris-HEPES-EDTA) buffer, 0.2 mL 2 M $MgCl_2$,10 mL 40% acrylamide solution (acrylamide: bis-acrylamide 29:1), 160 μL 25% Ammonium persulfate (APS), 40 μL TEMED, and 22 mL MilliQ water. To prepare 2 L of 5× THE buffer, dissolve in DEPC water 40 g Tris Base, 157 g HEPES, and 2 mL 0.5 M EDTA (pH 8.0). Bring to 2 L volume and filter to sterilize.
5. Combine 5 μL of transcription reaction with 5 μL 2× native gel loading buffer. 2× native gel loading buffer: 100 mM Tris-Cl pH 7.0, 20% (v/v) glycerol, 0.025% (w/v) Bromophenol Blue, 0.025% (w/v) Xylene Cyanol. Load 1–5 μL per lane. Run the native gel in 1× THEM buffer at 25 W (constant power) until the bromophenol blue dye migrates close to the bottom of the gel. 1× THEM buffer (1 L): 200 mL 5× THE buffer, 5 mL 2 M $MgCl_2$ (to 10 mM), and 795 mL MilliQ water. Keep the temperature of the native gels below 30 °C using fans and metal plates affixed to gel plates.
6. Following electrophoresis, stain the gel in ethidium bromide (EtBr) or SYBR Gold staining solution for 5 min and destain in deionized MilliQ water for another 5 min. Change water two times to reduce background staining. Image the gel using an appropriate gel-imaging system such as an ImageQuant 800 or Typhoon scanner.
7. The transcription reaction may generate some residual precipitation of magnesium pyrophosphate ($Mg_2P_2O_7$) despite the presence of inorganic pyrophosphatase in the reaction mixture, producing a cloudy appearance. Remove any precipitate by centrifugation at 9000 g for 5 min.
8. Use 10 kDa molecular weight cut-off (MWCO) concentrators to concentrate and desalt the supernatant. Centrifuge at 5500 × g for 30 min at 4 °C to reduce the volume by 80–90%. Dilute the sample back to the original volume with a low-salt buffer and repeat the centrifugation.

9. Recover the RNA sample from the concentrator and filter through a 0.2-micron filter to remove any residual precipitate in preparation for chromatographic separation. Avoid freeze–thaw cycles that may dissociate the RNA–RNA complex.

3.4 Purification of Large RNA–RNA Complexes by Size-Exclusion Chromatography (SEC)

1. After desalting and filtration, load the sample onto a size-exclusion column to purify the complex. RNA–RNA complex will be separated from the individual transcripts, abortive transcripts, RNase-cleaved RNA, DNA templates, protein enzymes including Taq DNAP, T7 RNAP and inorganic pyrophosphatase, NTPs, and other small molecules (Fig. 1). A Cytiva Superdex 200 Increase 10/300 GL (analytical) or HiLoad Superdex 200 16/600 (preparative) column is effective to separate the *B. subtilis* glyQS T-box discriminator–tRNA^{Gly} complex (144 nts, ~46 kDa, Kd ~ 2 μM) from free tRNA (75 nts, ~ 24 kDa) (Fig. 1b). Excess tRNA is used to saturate the discriminator, as the free discriminator does not separate well from the complex.
2. Equilibrate the column by running 2 column volumes (CV) of SEC buffer at a flow rate of 1 mL/min. SEC buffer: 10 mM Tris-HCl pH 7.4, 100 mM KCl, and 20 mM $MgCl_2$ (*see* **Note 2**).
3. Inject a concentrated RNA complex sample onto the column using an ÄKTA Pure system at the flow rate of 1 mL/min and continue with isocratic elution. (*see* **Note 3**) Reducing sample injection volume increases complex stability and improves separation resolution.
4. Collect the eluted samples in 0.5–1 mL fractions using a fraction collector into 96-well deep-well plates. Analyze relevant fractions by native and denaturing gel electrophoresis and pool appropriate fractions based on the chromatogram and analytical gels (Fig. 1c, d) (*see* **Note 4**). The sample is now ready for X-ray crystallographic and SAXS analyses.

3.5 Purification of Large RNA–RNA Complexes by Anion Exchange Chromatography

1. Anion exchange chromatography on a strong anion exchanger such as the Mono Q achieves superior resolutions than SEC. However, due to the extensive negative charge of larger RNAs, elution generally requires substantial ionic strength (up to 800 mM NaCl) which may dissociate some RNA–RNA complexes under study. The full-length T-box riboswitch and tRNA form extraordinarily stable complexes due to coordinated, multivalent base-pairing and coaxial stacking interactions (Fig. 2) [18]. This complex can withstand high ionic strengths and enable high-resolution separation of the 1:1 complex from the free T-box, which is not achievable using SEC as previously reported [26].

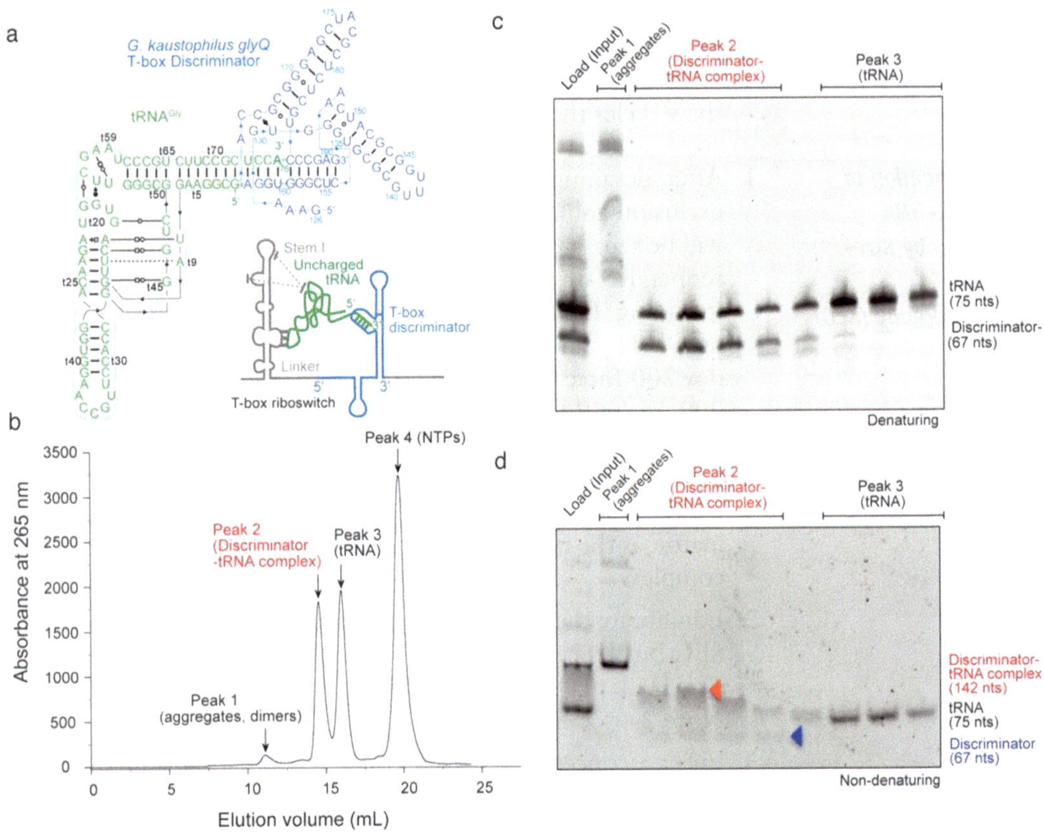

Fig. 1 Size-exclusion chromatography (SEC) purification of a T-box discriminator – tRNA^{Gly} complex. (**a**) Secondary structure of *Geobacillus kaustophilus glyQ* T-box Discriminator – tRNA^{Gly} complex informed by the co-crystal structure (PDB ID: 6PMO). Inset: cartoon scheme of a typical full-length glycine-specific T-box riboswitch bound to its cognate tRNA substrate. Dotted lines indicate stacking interactions between the apical region of the Stem I and tRNA elbow. (**b**) SEC profile of native, cotranscriptionally folded and assembled T-box Discriminator – tRNA^{Gly} complex. Main peaks are labeled, and their contents indicated. (**c**) Denaturing gel (10%) analysis of peak fractions of (**b**). Bands are annotated. (**d**) Nondenaturing gel (10%) analysis of the same peak fractions in (**b**, **c**)

2. Assemble a 5 mL cotranscription reaction by mixing 500 μL 10× T7 transcription buffer, 12.5 μL 2 M $MgCl_2$, 250 μL 100 mM each NTP, 67 μL T7 RNAP (prepared in house), 10 μL inorganic pyrophosphatase (1 Unit/μL), 1 mL full-length *B. subtilis glyQS* T-box PCR mixture, and 1 mL tRNA^{Gly} PCR mixture. Adjust final volume to 5 mL with DEPC water and incubate the reaction at 37 °C for 4–8 h. Desalt and filter the sample as described in Subheading 3.3.

3. Equilibrate a Mono Q column by running 5–10 column volumes of binding buffer at a flow rate of 0.5 mL/min. Binding buffer: 10 mM Tris-HCl pH 7.4, 100 mM KCl, and 20 mM $MgCl_2$.

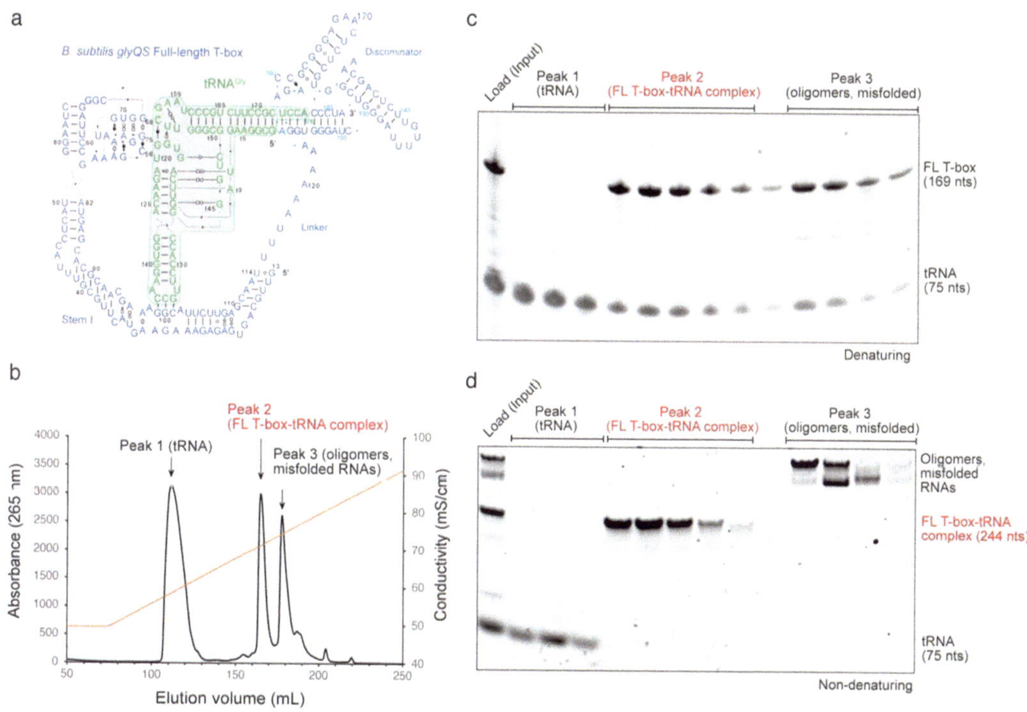

Fig. 2 Anion-exchange chromatography purification of a full-length T-box – tRNA^{Gly} complex. (**a**) Secondary structure of *Bacillus subtilis glyQS* full-length T-box– tRNA^{Gly} complex informed by two co-crystal structures (PDB ID: 6PMO; 4LCK) and cryo-EM structure (PDB ID: 6POM). (**b**) Mono Q profile of native, cotranscriptionally folded and assembled full-length T-box– tRNA^{Gly} complex. Main peaks are labeled, and their contents indicated. The conductivity trace representing the salt gradient is shown in orange. (**c**) Denaturing gel (10%) analysis of peak fractions of (**b**). Bands are annotated. (**d**) Nondenaturing gel (10%) analysis of the same peak fractions in (**b**, **c**)

4. Inject the RNA complex sample onto the Mono Q column. Unlike SEC, sample volumes do not significantly impact resolution on ion-exchange columns. Resolution is largely determined by the slope of the eluting salt gradient and the differences in molecular mass, shape, and total surface-exposed charge (Fig. 2b).
5. Elute the bound RNAs by applying a linear KCl gradient of 100 mM-1 M over 50–100 CV. While the tRNA elutes early at ~350–450 mM KCl, stoichiometric full-length T-box-tRNA elutes at ~600–700 mM KCl (Fig. 2b). Misfolded or aggregated T-box RNA and other oligomers elute later at >700 mM KCl, presumably due to higher overall negative charges that stem from more extended conformations, exposed phosphate backbones, or RNA oligomerization or aggregation.
6. Assess the contents of the fractions by native and denaturing gel electrophoresis (Fig. 2c, d). Pool appropriate fractions that

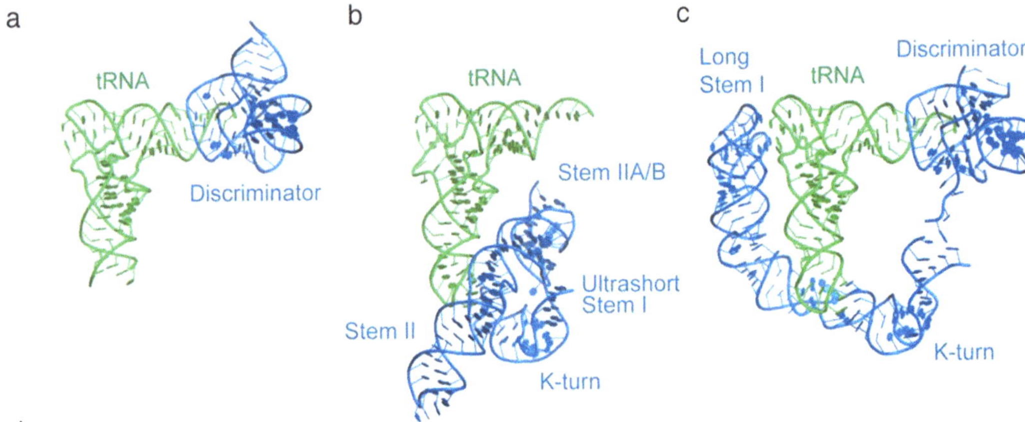

d

Species:	*G. kaustophilus*	*N. farcinica*	*B. subtilis*
T-box Domain:	Discriminator	Stem I-II-IIA/B	Full-length
PDB:	6PMO	6UFM	6POM
Resolution:	2.66 Å	2.82 Å	4.9 Å
Size:	46 kD	55 kD	78 kD
Method:	Size Exclusion	Size Exclusion	Anion Exchange
Column:	Superdex 200 Increase	Superdex 200 Increase	Mono Q

Fig. 3 Crystal and Cryo-EM structures solved using the protocol and corresponding method details. (**a**) Co-crystal structure of the *Geobacillus kaustophilus glyQ* T-box Discriminator – $tRNA^{Gly}$ complex (PDB ID: 6PMO). (**b**) Co-crystal structure of the *Nocardia farcinica ileS* T-box Stem I-Stem II domains in complex with the cognate $tRNA^{Ile}$ (PDB ID: 6UFM). (**c**) Cryo-EM structure of the *Bacillus subtilis glyQS* T-box full-length RNA (anti-terminated conformation) in complex with its cognate $tRNA^{Gly}$ (PDB ID: 6POM). (**d**) Characteristics of RNA samples and method parameters employed to solve the structures in (**a–c**)

contain the stoichiometric T-box-tRNA complex, desalt and concentrate using Amicon spin concentrators, filter through UltraFree filter units, concentrate further, and measure the concentration using a Nanodrop spectrophotometer prior to storage. The highly purified RNA complex sample is now ready for SAXS, Cryo-EM, or X-ray crystallographic analyses (Fig. 3).

7. Wash the column with 2 column volumes (CV) of 2 M NaCl, 4 CV of 0.25 M NaOH, 3 CV of 2 M NaCl, and 4 CV of DEPC water to remove all the bound molecules. Fill the column with 4 CV of 20% ethanol for storage.

4 Notes

1. tRNA acts as a folding chaperone for the T-box discriminator and full-length T-box. tRNA can be prepared either in advance as purified, refolded RNA added at the start of the in vitro transcription reaction or be transcribed and folded simultaneously with the T-box RNA.
2. For RNA–RNA complexes that depend on Mg^{2+} for stability, such as the T-box-tRNA complexes, it is important to avoid depriving the sample of Mg^{2+} at any step of the protocol. Once dissociated, some individual RNAs will misfold and can no longer reform the complex.
3. Due to the extensive negative charge of RNA backbone phosphates, RNAs bear much larger, multilayered hydration shells. As a result, RNA elutes much earlier from SEC columns, exhibiting apparent molecular weights that are 3–5 times larger than globular proteins.
4. Similar to proteins, the shape and flexibilities of RNAs impact their behavior on columns such as SEC in addition to their sizes. Correctly folded, functionally relevant RNA–RNA complexes are generally much more compact than their misfolded counterparts and are simultaneously larger in molecular mass. As a result, they frequently elute close to more extended individual RNA components. Thus, careful analysis of the stoichiometry of component RNAs in the complex peaks by biochemical analyses can be required to ensure sample homogeneity. Comparison of chromatograms of individual RNAs with their complexes is usually diagnostic of such phenomena. For example, free full-length T-box elutes from SEC columns at approximately the same volume as its tRNA complex, giving the false impression of a pure 1:1 complex [26].

Acknowledgments

We thank G. Piszczek and D. Wu for support with biophysical analyses, and A. R. Ferré-D'Amaré, K. Suddala, and C. Bou-Nader for discussions. This work was supported by the Intramural Research Program of the NIH, the National Institute of Diabetes and Digestive and Kidney Diseases (NIDDK) (ZIADK075136 to J. Z.), and an NIH Deputy Director for Intramural Research (DDIR) Challenge Award to J.Z. The authors declare no conflicts of interest.

References

1. Harris KA, Breaker RR (2018) Large noncoding RNAs in bacteria. Microbiol Spectr 6: 6.4.01.
2. Peselis A, Gao A, Serganov A (2015) Cooperativity, allostery and synergism in ligand binding to riboswitches. Biochimie 117:100–109
3. Reiter NJ, Osterman A, Torres-Larios A et al (2010) Structure of a bacterial ribonuclease P holoenzyme in complex with tRNA. Nature 468:784–789
4. Suslov NB, Dasgupta S, Huang H et al (2015) Crystal structure of the Varkud satellite ribozyme. Nat Chem Biol 11:840–846
5. Ferré-D'amaré AR, Scott WG (2010) Small self-cleaving ribozymes. In: Gesteland RF, Cech TR, Atkins JF (eds) The RNA world, 4th edn. Cold Spring Harbor Laboratory Press, Cold Spring Harbor
6. Korostelev A, Trakhanov S, Laurberg M et al (2006) Crystal structure of a 70S ribosome-tRNA complex reveals functional interactions and rearrangements. Cell 126:1065–1077
7. Ramakrishnan V (2002) Ribosome structure and the mechanism of translation. Cell 108: 557–572
8. Bou-Nader C, Zhang J (2020) Structural insights into RNA dimerization: motifs, interfaces and functions. Molecules 25:2881
9. Hood IV, Gordon JM, Bou-Nader C et al (2019) Crystal structure of an adenovirus virus-associated RNA. Nat Commun 10:2871
10. Zhang J, Ferre-D'amare AR (2014) Dramatic improvement of crystals of large RNAs by cation replacement and dehydration. Structure 22:1363–1371
11. Stagno JR, Liu Y, Bhandari YR et al (2017) Structures of riboswitch RNA reaction states by mix-and-inject XFEL serial crystallography. Nature 541:242–246
12. Somarowthu S, Legiewicz M, Chillon I et al (2015) HOTAIR forms an intricate and modular secondary structure. Mol Cell 58:353–361
13. Wickiser JK, Winkler WC, Breaker RR et al (2005) The speed of RNA transcription and metabolite binding kinetics operate an FMN riboswitch. Mol Cell 18:49–60
14. Zhang J, Lau MW, Ferré-D'amaré AR (2010) Ribozymes and riboswitches: modulation of RNA function by small molecules. Biochemistry 49:9123–9131
15. Zhang J, Landick R (2016) A two-way street: regulatory interplay between RNA polymerase and nascent RNA structure. Trends Biochem Sci 41:293–310
16. Wong TN, Sosnick TR, Pan T (2007) Folding of noncoding RNAs during transcription facilitated by pausing-induced nonnative structures. Proc Natl Acad Sci USA 104:17995–18000
17. Zhang J (2020) Unboxing the T-box riboswitches-A glimpse into multivalent and multimodal RNA-RNA interactions. Wiley Interdiscip Rev RNA 11:e1600
18. Li S, Su Z, Lehmann J et al (2019) Structural basis of amino acid surveillance by higher-order tRNA-mRNA interactions. Nat Struct Mol Biol 26:1094–1105
19. Suddala KC, Zhang J (2019) High-affinity recognition of specific tRNAs by an mRNA anticodon-binding groove. Nat Struct Mol Biol 26:1114–1122
20. Coleman TM, Wang G, Huang F (2004) Superior 5′ homogeneity of RNA from ATP-initiated transcription under the T7 phi 2.5 promoter. Nucleic Acids Res 32:e14
21. Martin CT, Coleman JE (1989) T7 RNA polymerase does not interact with the 5′-phosphate of the initiating nucleotide. Biochemistry 28: 2760–2762
22. Vasilyev N, Serganov A (2016) Preparation of short 5′-triphosphorylated oligoribonucleotides for crystallographic and biochemical studies. Methods Mol Biol 1320:11–20
23. Ferré-D'amaré AR, Doudna JA (1996) Use of cis- and trans-ribozymes to remove 5′ and 3′ heterogeneities from milligrams of *in vitro* transcribed RNA. Nucleic Acids Res 24:977–978
24. Xiao H, Murakami H, Suga H et al (2008) Structural basis of specific tRNA aminoacylation by a small *in vitro* selected ribozyme. Nature 454:358–361
25. Kao C, Zheng M, Rudisser S (1999) A simple and efficient method to reduce nontemplated nucleotide addition at the 3 terminus of RNAs transcribed by T7 RNA polymerase. RNA 5: 1268–1272
26. Zhang J, Ferré-D'amaré AR (2014) Direct evaluation of tRNA aminoacylation status by the T-box riboswitch using tRNA-mRNA stacking and steric readout. Mol Cell 55:148–155

Chapter 2

Probing RNA Structures and Interactions Using Fluorescence Lifetime Analyses

Jinwei Zhang

Abstract

Structural analyses of large, complex noncoding RNAs continue to lag behind their rapid discovery and functional descriptions. Site-specifically incorporated, minimally invasive fluorescent probes such as 2-aminopurine (2AP) and pyrrolo-cytosine (PyC) have provided essential complementary information about local RNA structure, conformational dynamics, and interactions. Here I describe a protocol that benchmarks and correlates local RNA conformations with their respective fluorescence lifetimes, as a general technique that confers key advantages over fluorescence intensity-based methods. The observation that fluorescence lifetimes are more sensitive to local structures than sequence contexts suggests broad utility across diverse RNA and ribonucleoprotein systems.

Key words RNA structure, Fluorescence, RNA–RNA interactions, Riboswitches, tRNA, T-box

1 Introduction

Emerging enabling technologies such as RNA-seq and Nanopore have transformed the landscape of RNA biology in the past two decades, and greatly accelerated the discovery and functional description of the noncoding genome. However, high-resolution structural analyses of biologically important noncoding RNAs and their protein complexes remain technically challenging. Effective analyses are hampered by RNA's intrinsic conformational heterogeneity and flexibility, highly charged, nonspecific and repetitive surfaces, and nonnative, solid-phase imaging environments that either densely pack RNAs in crystals or embed them in thin ice on Cryo-EM grids [1, 2]. Therefore, complementary solution-based methods are essential to corroborate and extend structural findings, and to provide insights into the dynamic property of the RNA structures, which frequently underlies their biological functions.

Jienyu Ding et al. (eds.), *RNA Structure and Dynamics*, Methods in Molecular Biology, vol. 2568,
https://doi.org/10.1007/978-1-0716-2687-0_2,

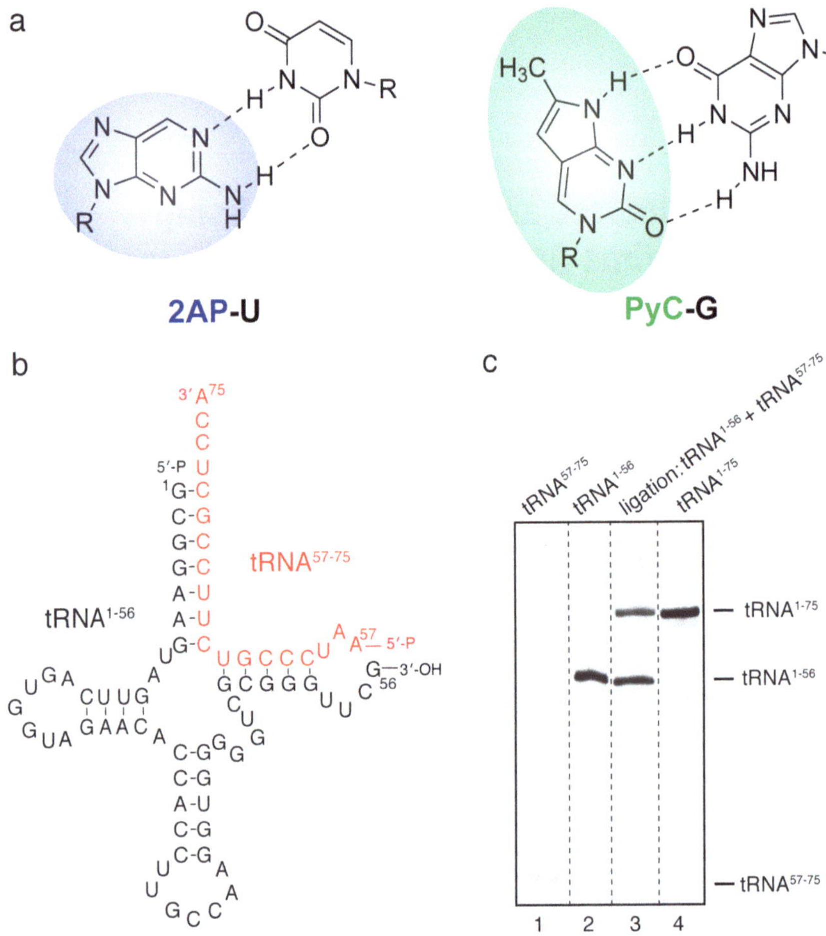

Fig. 1 Semi-enzymatic synthesis of 2AP-labeled tRNA. (**a**) Chemical structure of 2-aminopurine and pyrrolo-cytosine with their respective Watson-Crick base pairs. (**b**) General schematic of assembly and enzymatic ligation from two RNA fragments to produce full-length tRNAs that contain specific labels near their 3′ region. (**c**) Urea-PAGE analysis of the ligation reaction in (**b**). Note the lower band of tRNA$^{57-75}$ oligo stains weakly

Fluorescent nucleoside analogues such as 2-aminopurine (2AP) and pyrrolo-cytosine (PyC) have been widely used to probe nucleic acid structure and interactions, as they are capable of forming Watson-Crick base pairs, minimally perturb the geometry of local structures, and exhibit quantum yields that are sensitive to local environments [3–9] (Fig. 1). Most such applications are based on measurements of fluorescence intensity, whose precision and reproducibility are subject to a number of external variables including sample concentrations, incident light intensities, detector gain and sensitivities. By contrast, fluorescence lifetime is an intrinsic property of the fluorophore in a given chemical environment. It is generally independent of sample concentration, sample absorption, excitation light intensity, detector sensitivity, fluorescence emission intensity, and is unaffected by photobleaching. Nevertheless, it is still sensitive to temperature, polarity, and quenchers in solution. Fluorescence lifetime is reduced by Förster resonance energy transfer (FRET) and can quantitatively report FRET efficiency using a single variable that can be accurately measured and compared. Further, it can even be used where the fluorescence acceptor is not a fluorophore but a quencher, such as with molecular beacons.

Site-specific labeling of long RNA with fluorophores, especially at internal positions, remains a technical issue. While custom short RNA oligonucleotides under study can be chemically synthesized by several commercial sources, longer RNAs (> 60 nucleotides, or nts) generally require hybridization-based assembly of two or more RNA fragments with optional enzymatic ligation to seal nicks. Notably, recent technological developments, such as the position-selective labeling of RNA (PLOR) method, enables site-specific labeling of long RNAs by "walking" the RNA polymerase along the DNA template discontinuously [10]. PLOR operates by immobilizing the transcription elongation complex (TEC) on solid support, withholding one or more nucleotide triphosphates (NTP) in each cycle of transcript extension to allow partial extension, and iterative rounds of washing to supply only the appropriate subsets of NTPs [11, 12]. Multiple strategies for specific tRNA labeling have been developed [9, 13–15]. Here, we label the 3′-end of tRNAs with 2AP employing a flexible two-piece ligation scheme, where an unlabeled, in vitro transcribed 5′ fragment (nts 1–56) is enzymatically ligated with a partially complementary, chemically synthesized 3′ fragment (nts 57–75) bearing a terminal 2AP, forming the full-length tRNA [16, 17]. Similarly, we label a *Bacillus subtilis glyQS* full-length T-box riboswitch by annealing an in vitro transcribed 5′ fragment (nts 1–170) with a chemically synthesized 3′ fragment (nts 171–182) bearing an internal PyC.

In the method below, I describe the use of 2AP and PyC lifetime analyses to probe local RNA structural contexts, as well as RNA–RNA interactions that mediate the gene regulation by the T-box riboswitches. Using 2AP or PyC-containing benchmarking oligonucleotides derived from the tRNA and T-box riboswitch sequences, we demonstrate the general applicability of the method and consistency of lifetime values that correspond to structural contexts as opposed to sequence contexts. Critical insights from such lifetime measurements were instrumental in deriving correct structural models that were recently verified by crystallographic and Cryo-EM analyses [2, 17–19].

2 Materials

1. Custom oligonucleotides labeled with 2-aminopurine (2AP) or pyrrolocytosine (PyC) (IDT, Horizon, Trilink, etc.)
2. *Taq* DNA polymerase, 5000 U/mL (New England Biolabs).
3. T7 RNA polymerase, 50,000 U/mL (New England Biolabs).
4. Diethylpyrocarbonate (DEPC)-treated water.
5. RNA Binding Buffer: 50 mM HEPES-KOH, pH 7.0, 100 mM KCl, 20 mM $MgCl_2$, 5 mM tris (2-carboxyethyl) phosphine (TCEP).
6. EasyLife-LS system (Photon Technology International) or FluoroMax Plus Spectrofluorometer equipped with DeltaTime TCSPC (Time Correlated Single Photon Counting) module (Horiba Scientific).
7. LUDOX colloidal silica solution (Sigma-Aldrich).
8. T4 RNA ligase 1 (ssRNA ligase).
9. 2AP-triphosphate (2AP-TP) (Jena Bioscience).
10. Elutrap® Electroelution System (Cytiva).

3 Methods

3.1 Semi-enzymatic Synthesis of 2AP-Labeled tRNA

1. Perform in vitro transcription of tRNA 5′ fragment (1–56 nts) using T7 RNA Polymerase (RNAP). 5′ monophosphates are installed on tRNAs by "GMP priming," where a mixture of GMP (20 mM) and GTP (2 mM) is used instead of GTP, producing ~91% 5′ monophosphorylated tRNAs [20, 21]. Purify the desired RNA fragment ($tRNA^{1-56}$) by preparative denaturing Urea-PAGE, electroelution on an Elutrap device (Cytiva), washing with 1 M KCl and extensive exchange into DEPC-treated water, and store RNA samples at −20 ° C or − 80 ° C before use [22–24].
2. Ligate the $tRNA^{1-56}$ with an $tRNA^{57-75}$ oligonucleotide (IDT) bearing a 5′ monophosphate and 3′ 2AP modification in place of A75, using T4 RNA ligase 1 (ssRNA ligase), which seals the nick in the TΨC-loop with ~50% efficiency (Fig. 1). Gel-purify the ligated full-length $tRNA^{1-75(2AP75)}$ as described in **Step 1** (*see* **Note 1**).

3.2 Instrument Setup and Validation

1. On a fluorescence lifetime instrument, such as an EasyLife-LS system (Photon Technology International) used in this chapter, or a FluoroMax Plus Spectrofluorometer equipped with DeltaTime TCSPC (Time Correlated Single Photon Counting) module (Horiba Scientific), install a pulse excitation diode

or laser light source at 280–310 nm (for 2AP) or 340 nm (for PyC) and set emission monochromators or bandpass emission filters at 376/20 nm (for 2AP) or 448/20 nm (for PyC) (*see* **Note 2**).

2. Measure the Instrument Response Function (IRF, also known as "prompt") using a 0.1–1% LUDOX colloidal silica solution (Sigma). To measure short lifetimes (< 100 ps) accurately, laser sources that have shorter pulse widths and faster detectors (such as Hybrid Picosecond Photon Detectors, or HPPDs) is generally needed.
3. Verify the instrument setup by measuring the lifetime of a sample of free 2AP-triphosphate (2AP-TP). The expected average lifetime of 2AP-TP is ~8–10 ns in aqueous solution (Fig. 2a).

3.3 Benchmarking Fluorescence Lifetimes

1. Measure 2AP fluorescence lifetimes on a set of benchmark oligonucleotides that hybridize into known ssRNA and dsRNA structures at 5–10 μM (Fig. 2b). Less material would be required using TCSPC-capable instruments with enhanced sensitivity. A typical decay curve of 2AP fluorescence has four distinct components including a very short-lived species (< 100 ps) corresponding to a highly stacked conformation and a long-lived species (~10 ns) corresponding to a mobile, unstacked conformation [7, 8]. Depending on the specific environment of 2AP, 1–3 decaying species are often sufficient to satisfactorily fit the decay curve, producing reasonable χ^2 values (< 1.0–1.5). Collected traces are corrected by accounting for the IRF, normalized, and fit to minimal number of exponential decay components that produced acceptable χ^2 and random residuals, using Felix32 spectroscopy software (Photon Technology International), or DAS6 decay analysis software (Horiba) (*see* **Note 2**). Then a linear average lifetime is calculated using the fractional amplitudes and lifetimes of each component. As distinct decay curves and lifetime distributions can by chance produce very similar amplitude-weighted average lifetimes, it is important to examine the full decaying curves, in addition to the reduced average lifetime numeric values.
2. Correlate 2AP benchmark lifetimes with their respective structural contexts. Using different sets of 2AP benchmarks using distinct sequences, we generally obtained comparable lifetime values, which are more sensitive to local structural contexts than to sequence contexts (Fig. 2b–c). The extent of quenching is more dependent on stacking than base pairing, consistent with stacking with the nearest neighbor being the principal quenching mechanism [4, 7, 8]. Terminally located 2AP exhibit average lifetimes of ~1.5 and 4.0 ns, when paired with

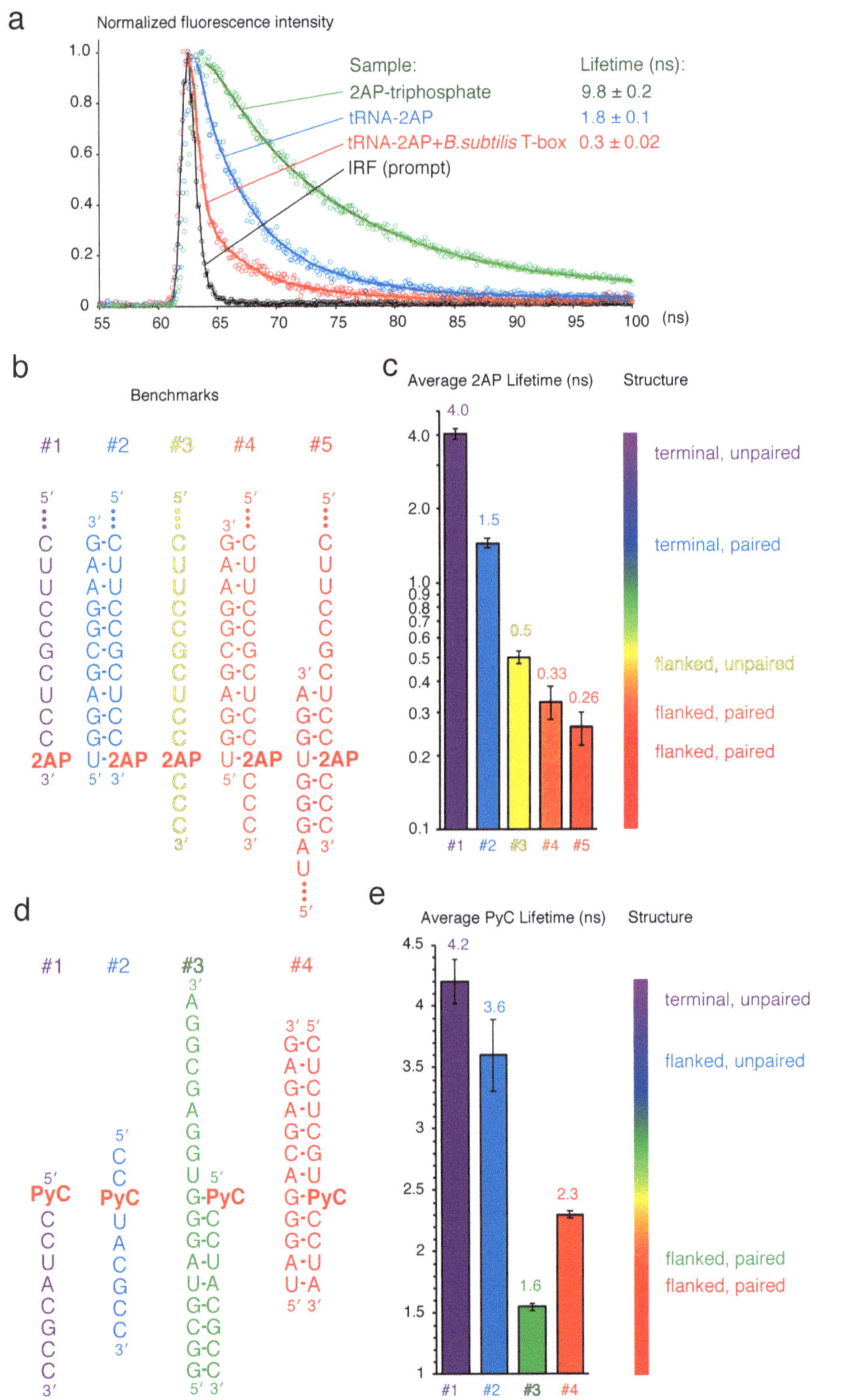

Fig. 2 2AP and PyC lifetime analyses and benchmarks. (**a**) Representative curves of the Instrument Response Function (IRF, or Prompt, black), 2AP fluorescence decay curves of 2AP-triphosphate (green), and 2AP-labeled

a uridine or unpaired, respectively (Fig. 2b–c, samples #1 and #2). This is likely due to the complete absence of stacking on one face of 2AP, as well as transient fraying of the duplex termini, which exposes both faces of 2AP to the solvent. Notably, an unpaired 2AP, when flanked by nucleobases on both sides, exhibited a much shorter lifetime of 0.5 ns (Fig. 2b–c, sample #3), even as both neighbors are pyrimidines that stack more weakly than purines. Importantly, the 2AP lifetime analysis is sensitive enough to distinguish between stacking on one face of 2AP versus stacking on both flanks, which can yield crucial information regarding local RNA structure, organization, or interaction [1]. When 2AP is located in the middle of a stable duplex, it exhibits a short lifetime of ~0.25 ns (Fig. 2b–c, sample #5). Interestingly, when a paired 2AP is located at the boundary between double-stranded RNA (dsRNA) and single-stranded RNA (ssRNA), it yields an intermediate lifetime of ~0.33 ns (Fig. 2b–c, sample #4). Together, these benchmarks demonstrate remarkable sensitivity and consistency of 2AP lifetimes as reporters of local RNA structure. It is important to note that sequence context of 2AP, especially adjacent guanosines in the same or opposing strand, can significantly impact 2AP lifetime through charge transfer [25, 26].

3. Correlate PyC benchmark lifetimes with their respective structural contexts. When the desired probe location is normally occupied by a pyrimidine rather than a purine, PyC can serve as a more appropriate probe than 2AP. Compared to 2AP, PyC has the advantage of longer excitation and emission wavelengths and longer lifetimes, reducing the technical requirements for instrumentation [13, 27]. The benchmarks of PyC lifetimes showed a pattern of sensitivity that is distinct from and more complex than that of 2AP (Fig. 2d–e). The data suggest that PyC lifetime appears more sensitive to base pairing than stacking with nearest nucleobases, which can either increase or decrease the lifetime of PyC depending on the structural and sequence context. This is consistent with theoretic calculations and previous reports that solvent shielding by neighboring nucleobases can enhance PyC fluorescence [5, 28, 29]. Nonetheless, the benchmarks suggest that PyC lifetime analyses effectively distinguish between ssRNA and dsRNA structural

Fig. 2 (continued) tRNA alone (blue) and in complex with a full-length T-box riboswitch (red). (**b**) 2AP-containing benchmark oligos derived from the *Bacillus subtilis glyQS* T-box riboswitch and $tRNA^{Gly}$ sequences. (**c**) Average 2AP lifetimes of the samples in (**b**) and a spectrum bar correlating local structure of 2AP with its lifetimes. (**d**) PyC-containing benchmark oligos derived from the *B. subtilis glyQS* T-box riboswitch. (**e**) PyC lifetimes of the samples in (**d**) and a spectrum bar correlating local structure of PyC with its lifetimes. All values reported throughout the chapter were measured on a PTI EasyLife-LS system (*see* **Note 2**)

contexts, but are unable to reliably distinguish between one-sided stacking and double-sided stacking, as 2AP can (*see* **Note 3**) [17].

4. (Optional) Examine the dependency of lifetimes on buffer solute conditions, especially divalent cation concentrations such as Mg2+. The conformations and dynamics of RNA structures can be modulated by the presence and concentrations of monovalent and divalent cations, in particular of Mg^{2+}(*see* **Note 4**) [30–32].

3.4 Measure Fluorescence Lifetimes of the Experimental Systems Under Study

1. With an empirical correlation between fluorescence lifetimes and local RNA structures established for a given sequence context on a particular instrument, actual experimental samples are ready to be measured. We illustrate this using samples of a free $tRNA^{2AP75}$, a free T-box riboswitch labeled by PyC, and their complexes.
2. Measure the lifetime of 2AP in free $tRNA^{2AP75}$. We observed a lifetime of ~1.8 ns, substantially shorter than the ~4.0 ns observed for a ssRNA oligo of the same sequence as the tRNA. This finding is consistent with a partially stacked conformation of the 3′-NCCA terminus [33, 34].
3. Assemble a native tRNA-*Bacillus subtilis glyQS* full-length T-box riboswitch complex sample by transcribing excess T-box RNA in the presence of purified $tRNA^{2AP75}$. Remove free labeled tRNA from the complexes using size-exclusion chromatography on a Superdex 200 Increase column. It is important to remove any labeled tRNA that is not in the complex, as they would obscure the lifetime analysis (*see* **Note 5**).
4. Measure the 2AP lifetime of the $tRNA^{2AP75}$ in the context of the T-box riboswitch–tRNA complex. We observed a highly quenched lifetime of 0.3 ns (Figs. 2a and 3a–c), which matches the lifetimes of 2AP in the middle of an RNA duplex from the benchmarks (0.26–0.33 ns; Fig. 2b–c). This finding provided a key clue to the structural organization of the T-box riboswitch–tRNA complex before structures were available (Fig. 3a–c). Namely, the tRNA 3′-adenosine, tA75, resides in a dsRNA duplex environment and is stacked on both sides [17]. As the tA75 is the terminal tRNA residue and stacks with tC74, then the other stacking must stem from the T-box. This critical insight suggests that the intermolecular helix formed between the tRNA 3′-NCCA and the T-box bulge stacks directly with Helix A1 of the T-box antiterminator, which allows the tRNA to stabilize the "ON" conformation of the T-box riboswitch (Fig. 3a–b). This exact structural arrangement has recently been validated by two independently solved co-crystal structures as well as a cryo-EM structure [2, 19, 35].

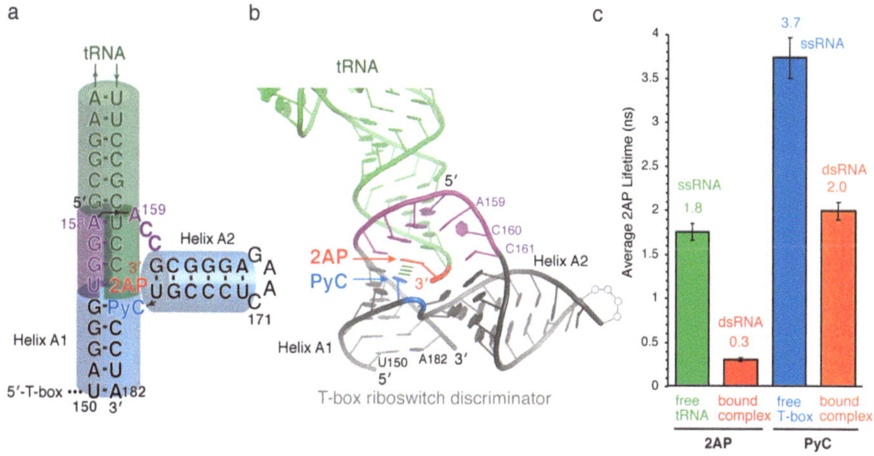

Fig. 3 2AP and PyC lifetime analyses of T-box riboswitch–tRNA complexes suggest structural model. (**a**) Secondary structure model of a *Bacillus subtilis* glyQS T-box riboswitch-tRNA complex proposed based on lifetime analyses. The locations of 2AP (red) and PyC (blue) are indicated. (**b**) Co-crystal structure of a *Geobacillus kaustophilus* T-box riboswitch discriminator domain–tRNA complex (PDB: 6PMO [2]). Stem III region is omitted for clarity. The locations of 2AP (red) and PyC (blue) are indicated. Numbering is the same as in (**a**). (**c**) 2AP lifetimes of $tRNA^{2AP75}$ either alone (green) or in complex with a full-length T-box riboswitch (red), and PyC lifetimes of $T\text{-}box171\text{–}182^{PyC178}$ RNA oligo either alone (blue) or in the context of a full-length T-box–tRNA complex (orange)

5. Similarly to 2AP, measure the lifetimes of PyC in a free RNA oligo corresponding to $T\text{-}box^{171–182}$, as well as when the oligo is annealed with the rest of the $T\text{-}box^{1–170}$ and in complex with unlabeled tRNA. Upon tRNA binding, the PyC lifetime was reduced from 3.7 ns to 2.0 ns (Fig. 3c), suggesting that the local environment of PyC changed from ssRNA to dsRNA, consistent with the 2AP findings and corroborated the structural model from the T-box side [17].

4 Notes

1. While 2AP and PyC substitutions rarely impact the biological function of the RNAs to significant extents, it is advisable to verify experimentally that these chemical modifications are indeed tolerated.
2. Absolute lifetime measurements depend significantly on the specific instrumentation used and its technical parameters, in particular, on the spectral width and pulse profile of the excitation light source. All values reported here were measured on a PTI EasyLife-LS system. For comparison, measurements of the same RNA samples on a Horiba FluoroMax Plus Spectrofluorometer equipped with DeltaTime TCSPC module generally produce longer lifetimes. Thus, extensive benchmarking on the specific instrument used for the experiments is required.

3. One key consideration for choosing between 2AP or PyC, when either nucleoside analogue is compatible with the biological system under study, is that 2AP fluorescence is more sensitive to nearest neighbor stacking than PyC. Using PyC as a stacking probe would generally require careful controls that take into account the specific sequence and structural contexts of PyC, with a particular emphasis on solvent shielding effects.
4. Since 2AP and PyC are most useful as a probe of local structure and direct interactions, it is advisable to be mindful that the global RNA conformations may change in unexpected ways when titrating divalent cations, ligands, or proteins, which may in turn cause drastic changes of the probe's local environment and fluorescence properties.
5. When measuring 2AP or PyC lifetimes in a binary or ternary complex, 1:1 stoichiometry is not required. The nonlabeled partner protein or RNA can be in excess, and should be, if the complex is unstable or has relatively low affinities. The excess binding partner will ensure that the labeled RNA remains in the complex, and will not interfere with lifetime measurements.

Acknowledgments

I thank G. Piszczek and D. Wu for support with biophysical analyses, and A. R. Ferré-D'Amaré, C. Bou-Nader, N. Baird, M. Lau, S. Li, K. Suddala, and K. Sapkota for discussions. This work was supported by the Intramural Research Program of the NIH, the National Institute of Diabetes and Digestive and Kidney Diseases (NIDDK) (ZIADK075136 to J.Z.), and an NIH Deputy Director for Intramural Research (DDIR) Challenge Award to J.Z. The author declares no conflicts of interests.

References

1. Zhang J, Ferré-D'amaré AR (2014) New molecular engineering approaches for crystallographic studies of large RNAs. Curr Opin Struct Biol 26:9–15
2. Li S, Su Z, Lehmann J et al (2019) Structural basis of amino acid surveillance by higher-order tRNA-mRNA interactions. Nat Struct Mol Biol 26:1094–1105
3. Sun A, Gasser C, Li F et al (2019) SAM-VI riboswitch structure and signature for ligand discrimination. Nat Commun 10:5728
4. Jean JM, Hall KB (2001) 2-Aminopurine fluorescence quenching and lifetimes: role of base stacking. Proc Natl Acad Sci USA 98:37–41
5. Hardman SJ, Thompson KC (2006) Influence of base stacking and hydrogen bonding on the fluorescence of 2-aminopurine and pyrrolocytosine in nucleic acids. Biochemistry 45:9145–9155
6. Lenz T, Bonnist EY, Pljevaljcic G et al (2007) 2-Aminopurine flipped into the active site of the adenine-specific DNA methyltransferase M.TaqI: crystal structures and time-resolved fluorescence. J Am Chem Soc 129:6240–6248
7. Jones AC, Neely RK (2015) 2-Aminopurine as a fluorescent probe of DNA conformation and the DNA-enzyme interface. Q Rev Biophys 48: 244–279

8. Xia T (2008) Taking femtosecond snapshots of RNA conformational dynamics and complexity. Curr Opin Chem Biol 12:604–611
9. Bou-Nader C, Muecksch, F, Brown, JB et al (2021) HIV-1 matrix-tRNA complex structure reveals basis for host control of Gag localization. Cell Host Microbe 29:1421–1436 e1427
10. Stagno JR, Liu Y, Bhandari YR et al (2017) Structures of riboswitch RNA reaction states by mix-and-inject XFEL serial crystallography. Nature 541:242–246
11. Liu Y, Holmstrom E, Yu P et al (2018) Incorporation of isotopic, fluorescent, and heavy-atom-modified nucleotides into RNAs by position-selective labeling of RNA. Nat Protoc 13:987–1005
12. Zhang X, Li M, Liu Y (2020) Optimization and characterization of position-selective labelling of RNA (PLOR) for diverse RNA and DNA sequences. RNA Biol 17:1009–1017
13. Zhang CM, Liu C, Christian T et al (2008) Pyrrolo-C as a molecular probe for monitoring conformations of the tRNA 3′ end. RNA 14: 2245–2253
14. Betteridge T, Liu H, Gamper H et al (2007) Fluorescent labeling of tRNAs for dynamics experiments. RNA 13:1594–1601
15. Masuda I, Igarashi T, Sakaguchi R et al (2017) A genetically encoded fluorescent tRNA is active in live-cell protein synthesis. Nucleic Acids Res 45:4081–4093
16. Beuning PJ, Musier-Forsyth K (1999) Transfer RNA recognition by aminoacyl-tRNA synthetases. Biopolymers 52:1–28
17. Zhang J, Ferré-D'amaré AR (2014) Direct evaluation of tRNA aminoacylation status by the T-box riboswitch using tRNA-mRNA stacking and steric readout. Mol Cell 55:148–155
18. Suddala KC, Zhang J (2019) High-affinity recognition of specific tRNAs by an mRNA anticodon-binding groove. Nat Struct Mol Biol 26:1114–1122
19. Battaglia RA, Grigg JC, Ke A (2019) Structural basis for tRNA decoding and aminoacylation sensing by T-box riboregulators. Nat Struct Mol Biol 26:1106–1113
20. Zhang J, Ferré-D'amaré AR (2013) Co-crystal structure of a T-box riboswitch stem I domain in complex with its cognate tRNA. Nature 500: 363–366
21. Martin CT, Coleman JE (1989) T7 RNA polymerase does not interact with the 5′-phosphate of the initiating nucleotide. Biochemistry 28: 2760–2762
22. Hood IV, Gordon JM, Bou-Nader C et al (2019) Crystal structure of an adenovirus virus-associated RNA. Nat Commun 10:2871
23. Klein DJ, Ferré-D'amaré AR (2006) Structural basis of *glmS* ribozyme activation by glucosamine-6-phosphate. Science 313:1752–1756
24. Milligan JF, Uhlenbeck OC (1989) Synthesis of small RNAs using T7 RNA polymerase. Methods Enzymol 180:51–62
25. Kelley SO, Barton JK (1999) Electron transfer between bases in double helical DNA. Science 283:375–381
26. O'Neill MA, Barton JK (2002) Effects of strand and directional asymmetry on base-base coupling and charge transfer in double-helical DNA. Proc Natl Acad Sci USA 99: 16543–16550
27. Tinsley RA, Walter NG (2006) Pyrrolo-C as a fluorescent probe for monitoring RNA secondary structure formation. RNA 12:522–529
28. Nguyen QL, Spata VA, Matsika S (2016) Photophysical properties of pyrrolocytosine, a cytosine fluorescent base analogue. Phys Chem Chem Phys 18:20189–20198
29. Hardman SJ, Botchway SW, Thompson KC (2008) Evidence for a nonbase stacking effect for the environment-sensitive fluorescent base pyrrolocytosine – comparison with 2-aminopurine. Photochem Photobiol 84: 1473–1479
30. Misra VK, Draper DE (2001) A thermodynamic framework for Mg2+ binding to RNA. Proc Natl Acad Sci USA 98:12456–12461
31. Grilley D, Soto AM, Draper DE (2006) Mg2+-RNA interaction free energies and their relationship to the folding of RNA tertiary structures. Proc Natl Acad Sci USA 103:14003–14008
32. Merriman DK, Yuan J, Shi H et al (2018) Increasing the length of poly-pyrimidine bulges broadens RNA conformational ensembles with minimal impact on stacking energetics. RNA 24:1363–1376
33. Limmer S, Hofmann HP, Ott G et al (1993) The 3′-terminal end (NCCA) of tRNA determines the structure and stability of the aminoacyl acceptor stem. Proc Natl Acad Sci USA 90: 6199–6202
34. Nagan MC, Beuning P, Musier-Forsyth K et al (2000) Importance of discriminator base stacking interactions: molecular dynamics analysis of A73 microhelix(Ala) variants. Nucleic Acids Res 28:2527–2534
35. Zhang J (2020) Unboxing the T-box riboswitches-A glimpse into multivalent and multimodal RNA-RNA interactions. Wiley Interdiscip Rev RNA 11:e1600

Chapter 3

Characterizing Fluorescence Properties of Turn-on RNA Aptamers

Robert J. Trachman III, Katie A. Link, Jay R. Knutson, and Adrian R. Ferré-D'Amaré

Abstract

Fluorescent RNA aptamers are tools for studying RNA localization and interactions in vivo. The photophysical properties of these in vitro selected RNAs should be characterized prior to cellular imaging experiments. Here, we describe the process of determining the fluorophore affinity, fluorescence enhancement, and fluorescence lifetime(s) of the Mango-III fluorescence turn-on aptamer. Parameters determined through these protocols will aid in establishing conditions for live-cell imaging.

Key words RNA aptamers, Fluorescence, Riboswitches, RNA imaging, RNA localization

1 Introduction

Analogous to fluorescent proteins, fluorescent RNA aptamers are genetically encodable tags useful for imaging RNA localization and interactions in live cells [1, 2]. Fluorescence turn-on aptamers have been in vitro selected to bind and enhance the fluorescence of small-molecule fluorophores. As such, characterization of binding affinity, fluorescence enhancement (Fl_E), and fluorescence lifetime are essential for successfully deploying these tags in live cells. Here we describe protocols for determining these three parameters starting with the determination of the dissociation constant (K_d), followed by fluorescence enhancement (Fl_E) measurement, and fluorescence lifetime. The sequence of measurements is logical, given that the K_d is used to determine Fl_E, and both K_d and Fl_E aid in determining conditions to measure fluorescence lifetime (*see* **Note 1**). In the following, we specifically describe the characteriza-

Robert J. Trachman III and Katie A. Link contributed equally to this work.

Jienyu Ding et al. (eds.), *RNA Structure and Dynamics*, Methods in Molecular Biology, vol. 2568,
https://doi.org/10.1007/978-1-0716-2687-0_3,

tion of the Mango-III aptamer [3, 4] in complex with the YO3-Biotin fluorophore (Mango-III-YO3-Biotin) [5]. However, the methods presented should be useful for measurement of the fluorescence properties of fluorescent RNA aptamers or fluorescent proteins that bind exogenous fluorophores, in general [6, 7].

2 Materials

All solutions were prepared using purified deionized water with a resistivity of 18.2 MΩ cm (*see* **Note 2**). All solutions were composed of reagents of >98% purity and filtered through 0.22 μm membranes.

2.1 *In Vitro* Transcription and RNA Purification

1. DreamTaq DNA polymerase (Thermo Fisher).
2. C1000 Thermocycler (BioRad).
3. 10x T7 transcription buffer: 700 mM Tris-HCl pH 7.0 (Fisher), 500 mM $MgCl_2$ (Sigma), 20 mM DTT (Sigma), 0.1% Triton-X (Sigma Aldrich).
4. 25 mM solution of each NTP (ATP, GTP, CTP, UTP) pH 7.0 (neutralized with NaOH; MP biomedical).
5. *E. coli* inorganic pyrophosphatase (Fisher).
6. 0.5 M EDTA pH 8.0 (pH adjusted with NaOH).
7. 99% Formamide.
8. 10x TBE buffer: 90 mM Tris-Borate (Fisher), 2 mM EDTA (Fisher).
9. 40% polyacrylamide solution (29:1; Acrylamide:Bis-Acrylamide) (Fisher).
10. TEMED (Fisher).
11. Ammonium persulfate (Fisher): 30% (w/v) solution.
12. Elutrap electroelution system (Whatman).
13. BT1 and BT2 membranes for Elutrap (Whatman).
14. 3 M KCl (Sigma).
15. RNA buffer: 20 mM MOPS-KOH (Sigma Aldrich) pH 7.0, 150 mM KCl (Sigma), 10 μM EDTA (Fisher).
16. Amicon Ultra-15 10 kDa M.W. cutoff centrifugal filter units (Millipore).
17. Ultrafree-MC centrifugal spin filters, 0.1 μm (Millipore).

2.2 Steady-State Fluorescence Measurements

1. YO3-Biotin [8].
2. Cary 300 Bio UV/Vis spectrophotometer (Varian).
3. NanoDrop 2000 spectrophotometer (Thermo Scientific).

4. Fluorimeter (PTI, or equivalently JYHoriba).
5. 0.1 cm quartz cuvette (Starna Cell Inc.).
6. KaliedaGraph (Synergy Software) or similar fitting program.

2.3 Fluorescence Lifetime Measurements

1. Thin quartz cuvettes (Precision Cells 52H, 0.700 mL capacity, 10 × 2 mm dual pathlength).
2. Silica, colloidal, 30% (w/v) suspension in water (Sigma-Aldrich).
3. Fluorescein or carboxyfluorescein, powder (Sigma or Molecular Probes).
4. Rhodamine 6G, powder (Exciton).
5. Sodium hydroxide, pellet (Sigma-Aldrich).
6. Ethanol, absolute (Sigma-Aldrich).
7. Coherent Mira 900-D Ti:sapphire laser (76 MHz, fs configuration) pulse picked with electro-optic modulator (Conoptics Model 350–160) and frequency doubled with a 2 mm BBO crystal (InRad).
8. Spectra Physics Vanguard 2000-HM532 NdYVO4 (80 MHz) pumping a Spectra Physics Model 3500 (4 MHz) cavity dumped dye laser.
9. JYH10 monochromator (8 nm bandwidth).
10. Polarizer in emission path, rotatable (Thor Labs).
11. Neutral density filters (or other attenuators) (Thor Labs).
12. Colored glass filters (Thor Labs).
13. Hamamatsu R3809U-50 MCP-PMT (Microchannel Plate – Photomultiplier Tube) detector.
14. 150 TCSPC (Time Correlated Single Photon Counting) module (Becker & Hickl) or "ADCAM" NIM modules for TCSPC (Ortec).

3 Methods

3.1 In Vitro Transcription

1. Perform polymerase chain reaction (PCR) by mixing DreamTaq components as described by the DreamTaq protocol (final volume, 1 mL). Add dNTP mixture to final concentration of 0.2 μM. Add forward and reverse primers, complementary to a T7 RNA polymerase start site and the ~17 complementary nucleotides of the 3′ end respectively, to a final concentration of 2 μM (*see* **Note 3**). Perform 20 cycles of PCR on a thermocycler with the annealing step at 58 °C.
2. Add 1 mL of PCR reaction product directly to the transcription reaction. Add 0.5 mL of 10x transcription buffer, 0.5 mL NTP solution, 1 unit/mL (final) inorganic pyrophosphatase, and

appropriate amount of T7 RNA polymerase (*see* **Note 4**). Add RNase-free H_2O to a final volume of 5 mL. Place in 37 °C water bath overnight (~10 h). Add 0.5 mL of 0.5 M EDTA pH 8.0 and vortex. Add 1.5 mL of 98% formamide and place on ice.

3. Prepare a large polyacrylamide gel capable of containing a single lane comb with dimensions 3.8 × 26.7 × 0.3 cm^3. Weigh 100 g of urea in 1 L beaker. Add 25 mL of 10x TBE and 94 mL of 40% polyacrylamide:bisacrylamide (29:1). Adjust volume to 250 mL with H_2O. Stir on magnetic hot plate until urea has dissolved (prepare the polyacrylamide slab gel assembly while urea is dissolving), but keep mixture at room temperature or below. Add 400 μL of 30% APS and 150 μL of TEMED while stirring. Pour into slab gel cassette. After polymerization is complete, remove comb and place gel on apparatus. Pre-run gel at a constant power of 70 W for approximately 1 h or until thermal equilibrium is reached.
4. Heat T7 transcription reaction-EDTA-formamide mixture to 60 °C for 5 min and load on prewarmed gel. Run gel at 70 W for approximately 3 h. Remove gel from apparatus and disassemble cassette. Cover gel in film wrap and place on a fluorescent TLC plate (*see* **Note 5**). In a dark room, briefly illuminate the gel with a handheld UV lamp and excise the band containing the RNA of interest with a clean razor or scalpel.
5. Elute the RNA from the gel using an Elutrap electroelution system. Place the RNA gel slices between the BT1 (macromolecule impermeable) and BT2 (macromolecule permeable) membranes in the gel cartridge. Fill all chambers with 1x TBE and run Elutrap overnight at 100 V. Extract RNA/TBE solution from chamber. Increase voltage to 200 V and run for an additional 2 h. Extract RNA/TBE solution.
6. Exchange RNA into RNA buffer using a 15 mL Amicon Ultra-15 10 kDa M.W. cutoff centrifugal filter unit. Place RNA solution from the Elutrap (~4 mL) into the centrifugal filter and add KCl to a final concentration of 1 M. Centrifuge at 5000 g until volume is reduced to ~0.25 mL. Add H_2O to 15 mL final volume and centrifuge at 5000 g until volume is reduced to ~0.25 mL. Add RNA buffer to 15 mL final and centrifuge at 5000 g until volume is reduced to ~0.25 mL (repeat two times for a total of three buffer exchanges). Filter the RNA solution with a 0.1 μm Ultrafree-MC centrifugal spin filter.
7. Determine the concentration of the RNA using a UV spectrophotometer. Dilute the RNA in H_2O and measure the absorbance at 260 nm and 280 nm. Measure the 260/280 nm ratio to help assess the purity of the sample. Determine the concentration of RNA using Beer's law (*see* **Note 6**).

3.2 Steady-State Fluorescence

3.2.1 Binding Affinity Measurement by Steady-State Fluorescence

1. Make an ~0.5μM solution of YO3-Biotin in RNA buffer and add to 1 cm pathlength quartz cuvette. Perform an absorbance scan from 240 nm to 680 nm on Cary 300 Bio spectrophotometer. Plot data to determine maximum absorbance wavelength (Abs_{max}).
2. Add the fluorophore to the RNA buffer so that the final concentration is no greater than the K_d (*see* **Note 7**). Add RNA aptamer to a final concentration of ~100 nM and place in 0.1 cm quartz cuvette. Exciting at Abs_{max}, perform a fluorescence emission scan from 590 nm (10 nm longer wavelength than Abs_{max}) to 660 nm, with 1 nm step size and 2 nm bandwidth. Plot data to determine maximum emission wavelength (Em_{max}).
3. Determine fluorescence excitation maximum (Ex_{max}) by measuring fluorescence at Em_{max} while scanning excitation wavelengths over 200 nm shorter wavelength range (410–610 nm). Plot data to determine Ex_{max}.
4. Using a NanoDrop Spectrophotometer, make three solutions of the RNA aptamer in RNA buffer ranging in concentration from 20 nM to 2μM. Anneal the RNA by heating to 95 °C for 3 min followed by cooling to room temperature in a sample holder on the bench (*see* **Note 8**).
5. Add 120μl of the fluorophore solution assembled in Subheading 3.2.2 and record an initial fluorescence reading by exciting at Ex_{max} and measuring Em_{max}.
6. Starting with 20 nM RNA aptamer solution, add 1 ml of RNA solution to the cuvette (mix thoroughly; *see* **Note 9**) and record fluorescence reading. Repeat with incrementally increasing volume of RNA aptamer added by 1μl (*i.e.*, 1μl, 2μl, 3μl) for 5 cycles. After 5 cycles, perform 5 cycles of the same titration series with 200 nM stock RNA aptamer solution, followed by the 2μM. The final volume in the cuvette will be 153μl.
7. Correct each fluorescence reading for dilution of the fluorophore. Multiply the fluorescence reading by the ratio of the current volume/initial volume.
8. Plot the dilution-corrected fluorescence emission reading (ordinates) vs. log10 concentration of RNA aptamer (abscissa). The data points should conform to a sigmoidal shape with two well-defined baselines. If not, adjust concentration of RNA titrants appropriately and repeat titration (Fig. 1).
9. Using KaliedaGraph or a similar program capable of performing nonlinear least-square analysis, fit the data to the Hill equation [9, 10] allowing the K_d and ligand stoichiometry to float. If the concentration of fluorophore at the mid-point of the titration is equal to or greater than the K_d, reduce the concentration of the fluorophore.

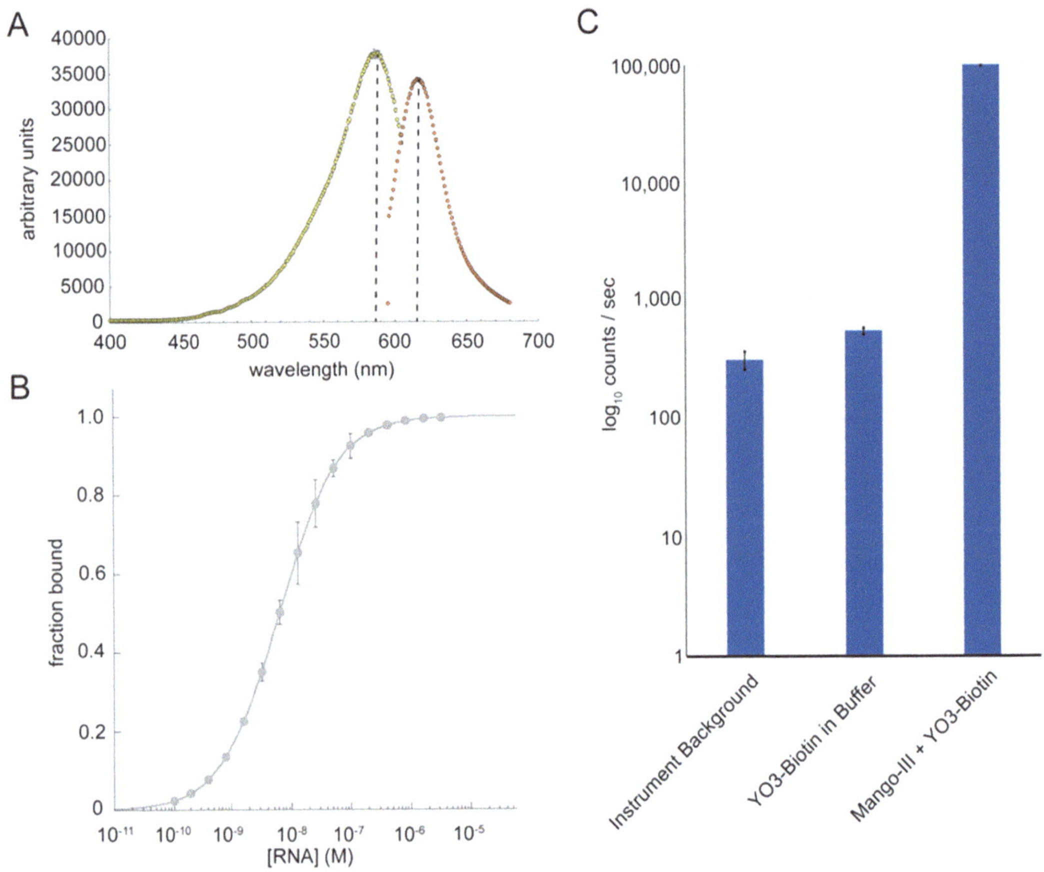

Fig. 1 Steady-state fluorescence characterization of Mango-III–YO3-Biotin. (**a**) Absorbance spectrum (yellow) and fluorescence emission spectrum (red) of the Mango-III–YO3-Biotin complex. (**b**) Binding isotherm for Mango-III-YO3-Biotin. Fluorescence measurements were performed at 20 °C at an excitation wavelength of 580 nm and an emission wavelength of 618 nm. Acquisitions were obtained with a 2 nm bandwidth and averaged for 2 s. (**c**) Fluorescence emission signal at 618 nm for the instrument alone, YO3-Biotin, and the Mango-III–YO3-Biotin complex

3.2.2 Fluorescence Enhancement Measurement by Steady-State Fluorescence

1. Determine the background signal of the fluorimeter (FlBK) by placing 120 μl of RNA buffer in 0.1 cm pathlength quartz cuvette and measure Em_{max}.
2. Add YO3-Biotin dissolved in RNA buffer to a final concentration of approximately 500 nM and measure fluorescence at Em_{max}. Fluorescence signal should be at least double the background signal. If not, add additional YO3-Biotin to cuvette.
3. Using the binding isotherm from Subheading 3.2.1, estimate the amount of RNA aptamer required to bind 99.99% of all YO3-Biotin in solution. Add prefolded RNA aptamer to the cuvette and take reading at Em_{max}. Make a correction for the dilution of YO3-Biotin fluorophore by multiplying Em_{max} by the quotient of the final volume and initial volume. Add an additional amount of RNA aptamer to ensure that no free YO3-Biotin is in solution.

3.3 Fluorescence Lifetime Measurement

1. Initialize instruments and lasers and allow them to come to thermal equilibrium to ensure maximum stability. Tune laser wavelength(s) to the excitation band of the fluorophore–aptamer complex (*see* **Note 10**). The Coherent Mira system is used for excitation bands below 490 nm and the Spectra Physics laser system is used for excitation bands between 560 and 590 nm.
2. The near-IR fs Mira is tunable from 700 to 980 nm. To produce a wavelength that overlaps with the excitation band of the fluorophore–aptamer complex, tune the Mira to double the desired wavelength. Place a beta barium borate (BBO) nonlinear doubling crystal in the optical path after the laser to double the frequency of the light (*see* **Note 11**). To double the frequency of the light the BBO should be oriented with its principal plane perpendicular to the polarization of the laser. Rotate the crystal until the doubled (second harmonic) beam is visible on fluorescent paper, after filtering the combination of IR and UV beams through a low-pass color filter. With the BBO, the operating wavelengths of the Mira ranges from ~350 to 490 nm (*see* **Note 12**).
3. The Mira repetition rate is too fast to collect the entire lifetime of the fluorophore. Use an electro-optic modulator (EOM) to pulse pick and provide sufficient time between pulses to collect the entire (*e.g.*, lifetime x 10) decay response (*see* **Note 13**).
4. Prepare a light scattering solution from the colloidal silica stock by diluting the stock 30% (w/v) suspension with pure deionized water (1:3). Add diluted suspension to one of the thin quartz cuvettes (*see* **Note 14**).
5. Prepare a slightly basic (pH ~8) aqueous solution of NaOH (<0.1 M) and fluorescein (~1μM). Add ~500μL to the 10 × 2.0 mm dual pathlength cuvette. Take the fluorescence measurement at or near the peak emission (512 nm). If the detector counting rate is saturated (*e.g.*, >3% of picked laser repetition), dilute further until the detector is no longer saturated.
6. Prepare a dilute (~1μM) rhodamine 6G solution in ethanol and add ~500μL to the 10 × 2.0 mm dual pathlength cuvette. Take the fluorescence measurement at or near the peak emission (~620 nm). If the detector rate is likely to "pile up" the detection counter (>3% of picked/dumped repetition), dilute further until the detector is no longer saturated.
7. Prepare room temperature-equilibrated samples of fluorophore-bound turn-on RNA aptamers in separate cuvettes and cover with a light blocking material until ready to use (*see* **Note 15**).

8. Measure the instrument response function (IRF) with the colloidal silica suspension. Put the colloidal silica cuvette in the sample holder. Attenuate the laser before the cuvette with neutral density filters (>100x) so that the detector is not saturated. In our configuration, both laser sources are vertically polarized so we set the polarizer after the sample and before the detector (emission polarizer) to 0° (vertical) as Rayleigh scattering is parallel to excitation. Take the transient of the silica suspension using the same excitation wavelengths used for the RNA aptamer sample (Rayleigh scattering, excitation=emission).
9. To set up the sample area for standards and fluorophore samples, set the emission polarizer to the "magic angle" (54.7°). This removes effects from rotational motion of the fluorophore. Use a blocking colored-glass filter (one that absorbs or reflects undesired excitation light but passes longer emission wavelengths) after the sample to prefilter fluorescence light passing through the monochromator to the PMT.
10. Take a fluorescence decay of the standards, rhodamine 6G when employing Spectra Physics dye laser (560–590 nm) and fluorescein when employing Coherent Mira (<490 nm) at the same wavelengths that will be used for the fluorophores. Excellent signal-to-noise ratios (SNR) are achieved when the peak counts in the decay exceed 10,000 and the total counts exceed one million (*see* **Note 16**).
11. Take fluorescence decay of RNA fluorophore complex(es) in question. For good SNR, the acquisition time typically targets >5000 photons/time channel at the peak.

3.4 Fluorescence Lifetime Analysis

1. Using the colloidal silica scattering solution, determine the "lamp" (IRF) profile. After loading the lamp profile, the standard (in our cases fluorescein or rhodamine 6G) should be analyzed with iterative reconvolution to optimize both the "color shift Q" and the choices for time channels where the fitting should begin and end. "Q" compensates for the small, typically <50 ps, difference in transit times between excitation and emission generated signals in the photon detector (transit time vs. wavelength shift).
2. Analyze data from fluorophore samples using reconvolution and nonlinear least squares optimization with Decay Fit (or similar) fitting program.
3. Using the optimized fixed parameters gleaned from standards (*e.g.*, Q-shift), fit the data for the sample to obtain minimal fit "cost parameters." A good fit will typically have a reduced χ^2 (*see* **Note 17**) value below 1.5 and ideally close to 1 (Fig. 2).
4. Once a good fit (a reduced χ^2 close to 1) has been found using the least number of fitting parameters necessary, find the 90%

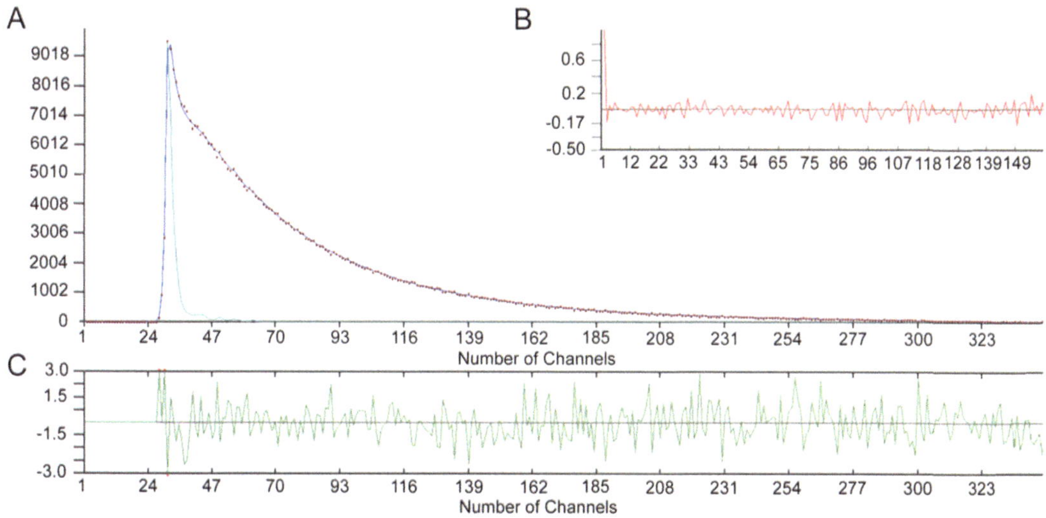

Fig. 2 Fluorescence lifetime measurement. (**a**) Fluorescence lifetime decay (red points) with fit (dark blue) and lamp (light blue), (**b**) autocorrelation, and (**c**) weighted residuals. The weighted residuals and their autocorrelation are used in addition to the reduced χ^2 to determine when the best possible fit is reached

confidence limits. In this case we will assume that two lifetimes ($\tau 1$ and $\tau 2$) were needed to get a good fit. To find the 90% confidence limits of $\tau 1$, adjust $\tau 1$ in the fit by fixing its value to push reduced $\chi^2 \pm 10\%$ (*see* **Note 18**) while other τ (lifetimes) and a (amplitudes) remain free. The change in $\tau 1$ required to surpass this limit will be the boundary for the 90% confidence limit. Repeat for $\tau 2$ (and any additional τ if more than 2 were needed).

4 Notes

1. Subpopulations of chromophore with ultrafast decay processes from probe flexure that steal potential quantum yield can be deduced from changes in the "radiative lifetime," defined as lifetime divided by quantum yield. Such discrepancies reveal "QSSQ" – ultrafast quenching [11, 12].
2. To minimize ribonuclease (RNase) contamination, deionized water can be treated with diethylpyrocarbonate (DEPC). To 4 liters of deionized water, add 1 ml of DEPC. Close the container and shake vigorously. Autoclave for 30 min. DEPC is a toxic, alkylating agent. Suitable precautions (personal protective equipment, fume hood) should be observed. Autoclaving destroys residual DEPC by converting it into ethanol and carbon dioxide.
3. In vitro transcription using T7 RNA polymerase often results in non-encoded nucleotide addition to the 3′ end of the

transcript. By adding 2′-O-methylated bases to the first two nucleotides of the reverse primer in a PCR reaction, incorporation of additional nucleotides to the 3′ end of the transcript can be reduced. Nucleotide addition can be further reduced by adding DMSO to the transcription reaction [13].

4. T7 RNA polymerase can be obtained from many sources including in-house expression and purification. As each preparation of this enzyme will vary in activity, care should be taken to ensure that the appropriate amount is used that maximizes RNA yield.
5. Fluorescent thin layer chromatography (TLC) plates can be used to detect nucleic acids, which absorb UV light and thus produce a shadow upon illumination. Care should be taken to limit the time in which RNA is exposed to a UV lamp so as to minimize photodamage [14].
6. Determining accurate RNA concentrations is critically dependent on the use of an accurate extinction coefficient. RNA extinction coefficients can be estimated from the RNA sequence; however, this may result in substantial error. A more accurate approach requires empirically determining the extinction coefficient through RNA hydrolysis. Using this approach, the RNA absorbance is measured before and after hydrolysis by an alkaline NaOH solution or heat [15]. Using the known sequence and extinction coefficients of each nucleotide, the extinction of the intact RNA can be determined.
7. Given the K_d is unknown, a low concentration of fluorophore that will result in a reasonably large signal when saturated with the fluorescent RNA-aptamer should be used.
8. Optimal annealing temperatures and times vary between RNA species. Annealing protocols should be tested prior to experimentation by varying temperature and time assessing folding efficiency on a native polyacrylamide gel or through size-exclusion chromatography.
9. Steady-state thermodynamic parameters can only be determined under reversible binding conditions, at equilibrium. It is important to ensure that the fluorescence signal is stable and that the system is not working towards equilibrium prior to taking a measurement. Using a fluorescence time-course experiment prior to titration should be used to ensure equilibrium [16].
10. A fixed diode laser can be used in lieu of a tunable laser if the laser emission of the diode laser overlaps with the excitation band of the fluorophore. The peak emission of the laser does not have to be exactly matched to the fluorophore's excitation max – there just needs to be overlap between the laser emission and excitation band of the fluorophore.

11. There are many nonlinear crystals in addition to a BBO crystal. A BBO was chosen because of their high damage threshold, their effective range (400-800 nm), and because they are widely available. Some other crystals that may work better for other wavelengths can be found in sources such as [17].
12. Always wear laser eye protection that attenuates beams of both wavelengths to final powers below 0.25 mW, in case a direct reflection of either beam to the eye accidentally occurs.
13. This reduced repetition rate can alternatively be achieved by cavity dumping a laser with an acousto-optic deflector.
14. A precise total volume of solution/suspension in the cuvettes is not critical; rather the solution/suspension must fill the cuvette enough so that the laser is below the meniscus and hitting neither the meniscus nor the bottom of the cuvette.
15. RNA must be well in excess over the fluorophore so as to reduce the lifetime fraction of unbound fluorophore state. The excess fraction of RNA required to minimize the unbound fluorophore can be calculated from a binding isotherm.
16. Other standards may be used. To choose another standard that is applicable, use a dye that is excited and emits at wavelengths similar to the fluorophore of interest.
17. Reduced χ^2 is the sum over all points of actual squared error divided by the expected squared error, then reduced (divided) by the "degrees of freedom" (number of points-number of parameters-1).
18. The choice of $\chi^2 \pm 10\%$ as the F-test limits was driven by using a 90% confidence interval with between 100-1000 channels to fit. More information about calculating confidence intervals can be found in [18, 19].

Acknowledgment

This work was supported by the intramural research program of the National Heart, Lung, and Blood Institute (NHLBI), National Institutes of Health. R.J.T. is the recipient of a K22 Career Transition Award from NHLBI.

References

1. Trachman RJ 3rd, Truong L, Ferré-D'Amaré AR (2017) Structural principles of fluorescent RNA aptamers. Trends Pharmacol Sci 38:928–939
2. Trachman RJ 3rd, Ferré-D'Amaré AR (2019) Tracking RNA with light: selection, structure, and design of fluorescence turn-on RNA aptamers. Q Rev Biophys 52:e8
3. Chen RF, Knutson JR, Ziffer H et al (1991) Fluorescence of tryptophan dipeptides: correlations with the rotamer model. Biochemistry 30:5184–5195

4. Xu J, Knutson JR (2009) Quasi-static self-quenching of Trp-X and X-Trp dipeptides in water: ultrafast fluorescence decay. J Phys Chem B 113:12084–12089
5. Autour A, Jeng SC, Cawte AD et al (2018) Fluorogenic RNA Mango aptamers for imaging small non-coding RNAs in mammalian cells. Nat Commun 9:656
6. Trachman RJ 3rd, Autour A, Jeng SCY et al (2019) Structure and functional reselection of the Mango-III fluorogenic RNA aptamer. Nat Chem Biol 15:472–479
7. Jeng SS, Trachman RJ 3rd, Weissenboek F et al (2020) Fluorogenic aptamers resolve the flexibility of RNA junctions using orientation-dependent FRET. RNA 27:433–444
8. Rodriguez EA, Tran GN, Gross LA et al (2016) A far-red fluorescent protein evolved from a cyanobacterial phycobiliprotein. Nat Methods 13:763–769
9. Yu D, Gustafson WC, Han C et al (2014) An improved monomeric infrared fluorescent protein for neuronal and tumour brain imaging. Nat Commun 5:3626
10. Jepsen MDE, Sparvath SM, Nielsen TB et al (2018) Development of a genetically encodable FRET system using fluorescent RNA aptamers. Nat Commun 9:18
11. Weiss JN (1997) The hill equation revisited: uses and misuses. FASEB J 11:835–841
12. Hill AV (1909) The mode of action of nicotine and curari, determined by the form of the contraction curve and the method of temperature coefficients. J Physiol 39:361–373
13. Helmling C, Keyhani S, Sochor F et al (2015) Rapid NMR screening of RNA secondary structure and binding. J Biomol NMR 63:67–76
14. Kladwang W, Hum J, Das R (2012) Ultraviolet shadowing of RNA can cause significant chemical damage in seconds. Sci Rep 2:517
15. Wilson SC, Cohen DT, Wang XC et al (2014) A neutral pH thermal hydrolysis method for quantification of structured RNAs. RNA 20: 1153–1160
16. Jarmoskaite I, Alsadhan I, Vaidyanathan PP et al (2020) How to measure and evaluate binding affinities. Elife 9:e57264
17. Nikogosyan DN (2005) Nonlinear optical crystals: a complete survey. Springer, New York
18. Johnson ML, Faunt LM (1992) Parameter estimation by least-squares methods. Methods Enzymol 210:1–37
19. Straume M, Johnson ML (1992) Analysis of residuals: criteria for determining goodness-of-fit. Methods Enzymol 210:87–105

Chapter 4

Probing Transient Riboswitch Structures via Single Molecule Accessibility Analysis

Robb Welty, Andreas Schmidt, and Nils G. Walter

Abstract

Riboswitches are a class of RNA motifs in the untranslated regions of bacterial messenger RNAs (mRNAs) that can adopt different conformations to regulate gene expression. The binding of specific small molecule or ion ligands, or other RNAs, influences the conformation the riboswitch adopts. Single Molecule Kinetic Analysis of RNA Transient Structure (SiM-KARTS) offers an approach for probing this structural isomerization, or conformational switching, at the level of single mRNA molecules. SiM-KARTS utilizes fluorescently labeled, short, sequence-complementary DNA or RNA oligonucleotide probes that transiently access a specific RNA conformation over another. Binding and dissociation to a surface-immobilized target RNA of arbitrary length are monitored by Total Internal Reflection Fluorescence Microscopy (TIRFM) and quantitatively analyzed, via spike train and burst detection, to elucidate the rate constants of isomerization, revealing mechanistic insights into riboswitching.

Key words Bacterial gene regulation, Conformational dynamics, Riboswitch, RNA folding, Single molecule fluorescence microscopy

1 Introduction

RNAs are dynamic biomolecules that not only encode genetic information but also modulate many biological processes. RNA molecules are able to interconvert dynamically between complex structures with a range of functionalities (e.g., catalysis, gene regulation), making them an important class of biomolecules to study for understanding biology [1–3].

Riboswitches, regulatory RNA elements found in the 5′ untranslated region (5′ UTR) of some prokaryotic messenger RNAs (mRNAs), adopt different structural ensembles as a function of cellular conditions. These conformations can either promote or inhibit the expression of the gene(s) in the coding region of the

Robb Welty and Andreas Schmidt contributed equally to this work.

Jienyu Ding et al. (eds.), *RNA Structure and Dynamics*, Methods in Molecular Biology, vol. 2568,
https://doi.org/10.1007/978-1-0716-2687-0_4,

hosting mRNA (i.e., acting in "cis"). Small molecule ligands preferentially bind to one conformation (*via* a highly specific aptamer domain), thereby stabilizing it. This feature allows for the concentration of the ligand to act as a rheostat that dynamically enhances the presence of a specific RNA conformation, in turn modulating gene expression [4]. This dynamic structure–function relationship helps shape prokaryotic gene expression profiles [5].

Single molecule approaches can detect and quantify individual biomolecular events. A subset of these techniques are suitable to probe the dynamic nature of riboswitch structure from which its modulation of protein expression can be inferred [6–10].

This chapter provides a protocol (focusing on probe design, reconstitution, immobilization, data acquisition, and analysis) for SiM-KARTS (Single Molecule Kinetic Analysis of RNA Transient Structure), whereby riboswitch dynamics are probed, in real time, without the limitations associated with direct labeling of target RNA (Fig. 1) [11, 12]. In brief, by hybridizing a biotinylated nucleic acid capture probe (CP) and a fluorescently labeled nucleic acid localization probe (LP) to distal ends of the riboswitch hosting mRNA of interest, surface immobilization and localization are enabled, respectively (Fig. 1). Next, a transiently binding, fluorescently labeled nucleic acid, the fluorescence probe (FP), is added in excess. It is designed to preferentially bind to one of the riboswitch conformations, whereas it is occluded from any other (Fig. 1b). The transient binding and dissociation of the FP is recorded by Total Internal Reflection Fluorescence Microscopy (TIRFM) [11]. The repeated binding events are recorded as spike trains in the fluorescence intensity originating from an individual riboswitch on the surface as a function of time (Figs. 2 and 3). To discriminate between LP and FP, the probes are labeled with two different fluorophores that have easily distinguishable spectroscopic properties, typically emission wavelength (Fig. 1b) [11]. This strategy permits the extraction of distinct kinetic patterns, or temporal fingerprints, of the binding of FP to a specific region of target RNA that can be used to discriminate between, and measure the interconversion of, multiple conformations of an embedded riboswitch. SiM-KARTS can be applied under potentially many buffer and biological matrix conditions across a wide range of timescales [10–13].

The current protocol is based on an investigation of a 7-aminomethyl-7-deazaguanine ($preQ_1$)-sensing riboswitch [12]. The $preQ_1$ riboswitch from *Thermoanaerobacter tengcongensis* (*Tte*) regulates gene expression by forming a pseudoknot that partially masks its Shine-Dalgarno, or SD, sequence [14, 15]. In the absence of $preQ_1$, the SD sequence is more readily available to hybridize to the anti-SD sequence of 16S ribosomal RNA, resulting in the translation of two downstream proteins, TTE_RS07450 and TTE_RS07445, involved in cellular $preQ_1$ metabolism [12].

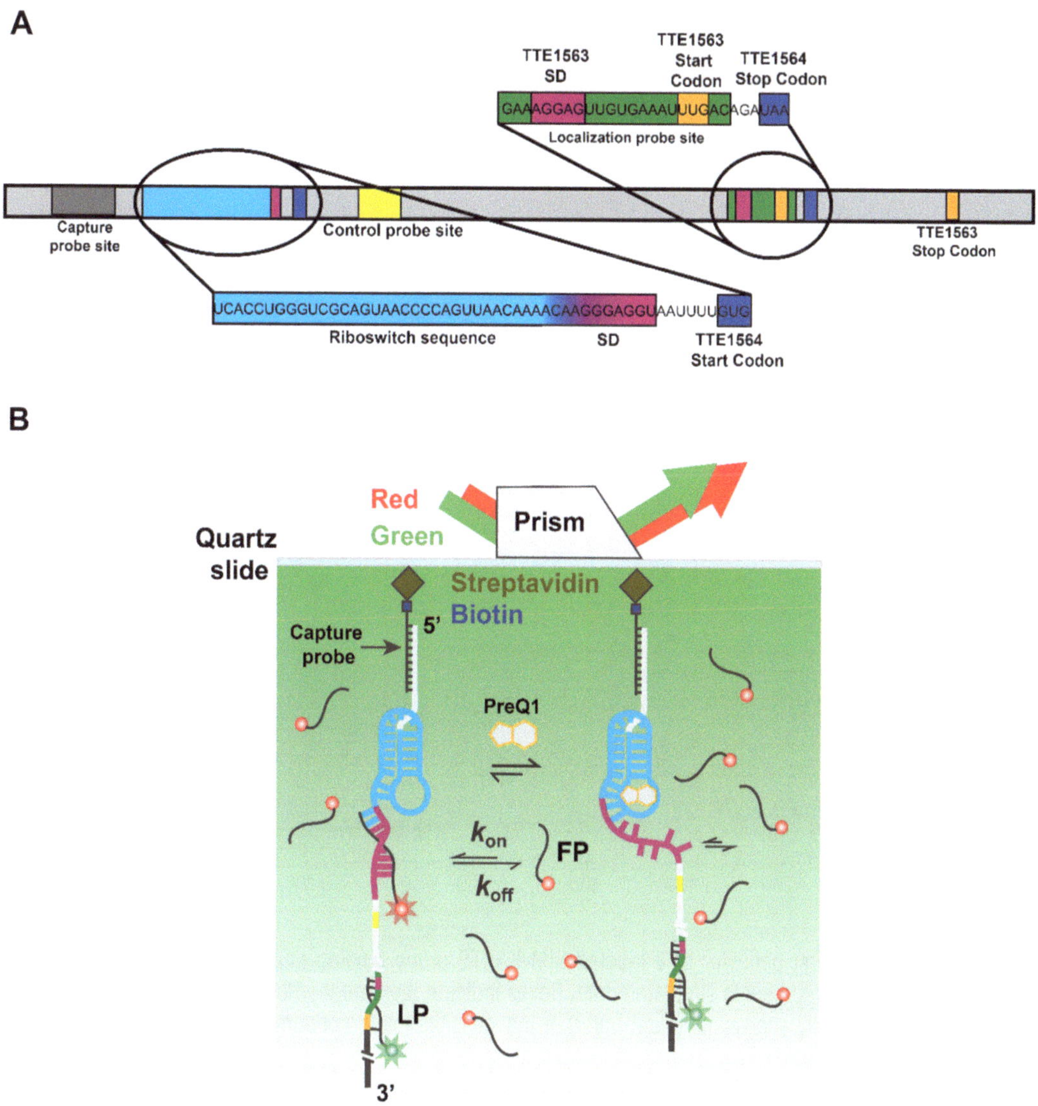

Fig. 1 SiM-KARTS measurements of an mRNA with embedded preQ$_1$ riboswitch. (**a**) The preQ$_1$ riboswitch containing mRNA. The transcript is immobilized onto the microscope slide after hybridization of the biotinylated capture probe (CP) to the corresponding recognition site on the target. The distally hybridized localization probe (LP) allows for the localization of single immobilized mRNA molecules. Transiently binding fluorescence probe (FP, here against the Shine-Dalgarno (SD) sequence of the TTE1564 protein) senses structural changes of the riboswitch. (**b**) Experimental schematic of the SiM-KARTS assay using a prism-based TIRFM setup. After the RNA complex is immobilized onto a microscope slide via streptavidin and biotinylated BSA (biotin-BSA, not shown for simplicity), it is illuminated with a 532 nm wavelength laser (exciting the LP) to identify the RNA complex. A red laser (638 nm) is used to excite the FP, whose repeated binding to the Shine-Dalgarno (SD) sequence results in spikes of fluorescence. In the presence of preQ$_1$, the accessibility of the binding site of the FP changes

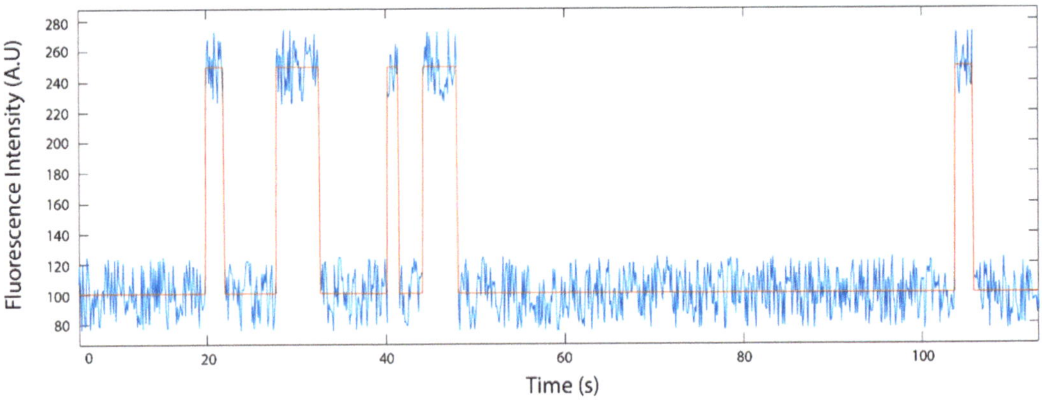

Fig. 2 Example single molecule fluorescence trace (blue), idealized here by intensity thresholding (orange)

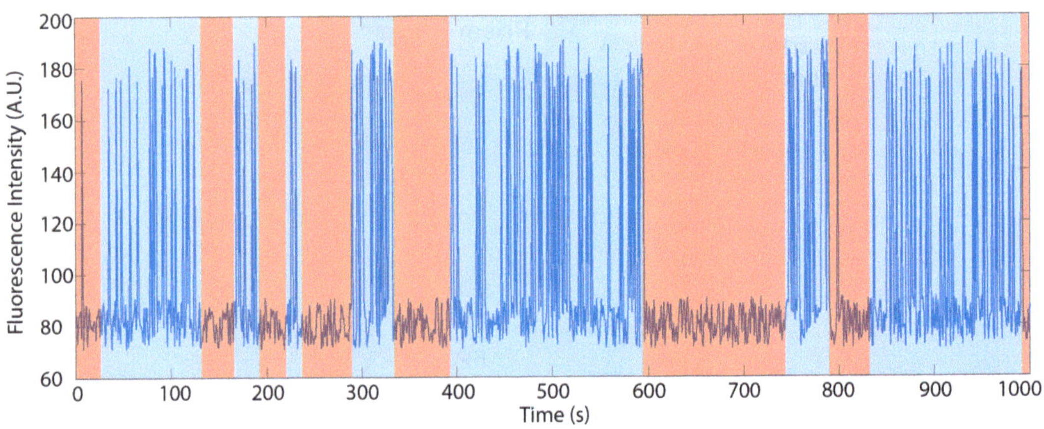

Fig. 3 Extraction of burst behavior of a labeled SiM-KARTS probe binding and dissociating via spike train analysis. The blue line represents simulated data. Boxes indicate the time the RNA is in a conformation where the binding site is readily accessible (blue; burst) or inaccessible (pink; nonburst)

SiM-KARTS interrogates the conformational changes of a target system via binding and dissociation of an FP. It can, in principle, probe any nucleic acid that isomerizes between different secondary structures so long as: (i) there is a portion of the target that alternates between being single- and double-stranded forms that can accommodate binding a short (typically <10 nucleotide complementarity) nucleic acid labeled with a fluorescent dye; (ii) the binding events are long enough to be resolved by single molecule TIRFM (typically ~100 ms or longer desired); and iii) the probe generates a distinct, site-specific kinetic fingerprint of binding and dissociation events that depends on the RNA conformation. Of note, targeting a conformational change provides an intrinsic control for any perturbation the FP may have on the structure probed, since only the relative (ratiometric) change is evaluated as a readout of the underlying conformational equilibrium.

2 Materials

The materials presented here describe the representative experiments in [12]. To generalize this materials section, the reader will want to replace the ligand, imaging buffer, mRNA transcript, and probes to pertain to the specific riboswitch or RNA under study. All solutions should be prepared in doubly deionized water (ddH_2O), obtained by further purifying deionized water to a resistivity of 18 MΩ cm, and using the highest purity of reagents available.

2.1 Ligand 7-aminomethyl-7-deazaguanine ($preQ_1$)

$PreQ_1$ can be synthesized according to [16] or simply purchased from Sigma Aldrich (Cat. Number: SML0807). Care needs to be taken to dissolve the $preQ_1$ into a high stock concentration (1 mM) since it is multi-protic, and the pH needs to be adjusted to neutrality.

2.2 Riboswitch Containing mRNA Transcript

The mRNA transcript containing the 5′ UTR riboswitch is typically generated by in vitro transcription [12, 17].

2.3 Biotinylated Bovine Serum Albumin (1 mg/ml)

Biotin-BSA can be purchased from diverse commercial sources, such as Sigma Aldrich (CAS Number 9048-46-8).

2.4 T50 Buffer

10 mM Tris-Cl, pH 8.0, 50 mM NaCl.

2.5 Oxygen Scavenging Stock Solution with TROLOX (OSST)

Each component is stored individually and combined immediately prior to usage; the mixture can be stored for up to 2 h:

1. Protocatechuate 3,4-dioxygenase (PCD), 1 μM in 100 mM Tris-HCl, pH 8.0, 50 mM KCl, 1 mM EDTA, 50% (v/v) glycerol. Stored at −20 °C. PCD is obtainable from Sigma Aldrich, CAS number 9029-47-4.
2. Protocatechuic acid (PCA), 100 mM dissolved in ddH_2O, adjusted with 5 M KOH to pH 8.3. Stored at −20 °C. PCA is obtainable from Sigma Aldrich, CAS number 99-50-3.
3. TROLOX, 100 mM dissolved in ddH_2O, adjusted with 5 M KOH to ~pH 10, filtered with 0.2 μm filter and stored at −20 °C (*see* **Note 1**). TROLOX is obtainable from Fisher Scientific, CAS number 53188-071.

2.6 2x Reconstitution Buffer

100 mM Tris-HCl, pH 7.5 at 25 °C, 1200 mM NaCl and 40 mM $MgCl_2$.

2.7 Imaging Buffer

1x reconstitution buffer supplemented with OSST (5 mM PCA, 50 nM PCD and 2 mM TROLOX).

2.8 Preparation of Slides with Microfluidic Channel

Protocols for preparation of a microfluidic channel on a slide can be found in [9, 11]. In brief, a quartz slide (G. Finkenbeiner, 1″ × 3″ × 1 mm microscope slides) has two 1 mm holes drilled in it approximately 1 inch apart. The flow channel is then created by placing double-sided adhesive tape (3 M catalogue no. 3136) to form a thin channel that starts and ends at the holes. A microscope coverslip is attached to the adhesive tape, completing the flow channel. During the SIM-KARTS experiment, fluids can be passed through the channel by tubing fixed to the holes (typically with epoxy) or via wicking action.

2.9 Oligonucleotides

Oligonucleotide probes are obtainable from, e.g., Integrated DNA Technologies. TYE563 can be substituted with Cy3. For a detailed description of probe design, *see* **Notes 2–6**. In the following, the probes are written in 5′ to 3′ direction.

1. Capture probe (CP): GCCTCTTAGCAACTTGTAGTAG GAGTTCCAAAAAAAAAA-biotin, which acts as an immobilization anchor of the RNA transcript onto the microscope slide.
2. Localization probe (LP): TYE563 – +GT + CAAATTT +CA + CAA + CT + C + CTTT+C, where a preceding "+" indicates a locked nucleic acid (LNA) nucleotide, which increases the hybridization stability; the LP serves to identify single target RNA molecules.
3. Fluorescent probe (FP): Cy5-GAUCACCUCCUU, which serves as an anti-SD probe to read out conformational changes of the riboswitch containing mRNA.
4. Control probe: Cy5-GCAACAAGAGC; acts as an internal control for specific structural changes of preQ1 riboswitch containing mRNA due to preQ1; ligand addition should not affect its binding behavior, in contrast to the FP's behavior.

2.10 Microscope and Hardware

A microscope capable of objective, or prism, TIRFM is needed. For details on model instruments, please see recent reviews [18, 19]. In short, our setup consists of two lasers, a Venti 532 (Laser Quantum) and an Obis 637 (Coherent), to excite the LP and FP, respectively; power of laser beams range between 50 and 100 mW over an area of typically 1.76 mm^2. Sample is loaded into a custom quartz microfluidic flow cell [18], which is mounted on an inverted fluorescence microscope (IX71 Olympus). The fluorescence emission from the LP and FP are split with a dichroic mirror and laser scatter is removed via bandpass filters. Fluorescence is recorded with a sCMOS (scientific complementary metal-oxide-semiconductor) camera (ORCA 3, Hamamatsu).

2.11 Software

Analysis of single molecule movies requires software for creating single molecule data traces. Multiple such programs are available for download (e.g., SPARTAN [20]), MASHFRET [21]; we use homebuilt scripts. The only requisite is that the fluorescence data can be exported in a comma separated values (CSV) format for processing with our SiM-KARTS pipeline. The SiM-KARTS burst detection algorithm (https://github.com/RobbWelty/SiMKARTS) requires MATLAB (version 6.5 or later with the statistical toolbox). However, the process is generalizable and can be rewritten into a different language if desired.

3 Methods

3.1 Preparation of RNA Complex for SiM-KARTS

1. Denature riboswitch containing mRNA transcript, LP and CP (2 nM each) at 70 °C in the presence of 1x reconstitution buffer for typically 2 min.
2. Allow the reaction to cool to room temperature (~21 °C) over 20 min in the presence (≥10 μM for saturation in the case of preQ1) or absence of ligand.
3. Dilute the reaction mixture to 40 pM in >200 μL using the same buffer in the presence or absence of ligand and supplemented with a 12.5-fold excess of both LP and CP to ensure the complex stays intact upon dilution. The complex can be chilled on ice. (*see* **Note** 7.)

3.2 Immobilization of Reconstituted preQ$_1$ Riboswitch Containing mRNA Complex onto Microscope Slides

All binding and washing steps are typically performed at room temperature.

1. Flow 200 μL of biotin-BSA (1 mg/ml) through the microfluidic channel and allow to bind for 10 min. (This step passivates the glass surface as well as introduces anchoring moieties for the mRNA complex.)
2. Wash unbound biotin-BSA with 200 μL of T50 buffer.
3. Flow through 0.2 mg/mL streptavidin and wait for 5 min to ensure saturation.
4. Flow through 200 μL of 1x reconstitution buffer to wash away unbound excess streptavidin.
5. Flow through 200 μL diluted and reconstituted riboswitch transcript complex and wait for 5 min to ensure saturation.
6. Flow through 200 μL of 1x reconstitution buffer to wash away unbound riboswitch transcript complex.
7. Flow through 200 μL of FP (50 nM) in imaging buffer, and wait 5 min for equilibration (*see* **Note 8**).
8. Start imaging.

3.3 Data Acquisition

1. Start image acquisition with both lasers that illuminate both the CP and LP. *See* **Note 9** for recommendations on hardware settings.
2. (Optional) Turn off the laser illuminating the LP after some number (e.g., 100) frames to reduce background from scattered laser illumination.
3. After acquisition, move the microscope slide to a new field of view and record another movie.
4. Repeat **step 4** until you have produced the desired number of movies (typically 5–10 movies are sufficient to generate at least 200 single molecule traces, but this will depend heavily on the biochemical and microscope system used).

3.4 Data Analysis

After the movies have been taken, the fluorescence intensities of the FPs must be extracted for the time course of the movie. There are many programs available for this purpose, including SPARTAN [20], MASH FRET [21], and multiple proprietary homebuilt scripts. To complete SiM-KARTS analysis, using the tools provided in this chapter, the only requisite is that the fluorescence intensity for single molecules over the time of the movie, called traces, must be in a CSV format with the columns corresponding to individual LP positions and the rows corresponding to the time the intensity was collected. To increase the accuracy of spot selection, we recommend mapping the coordinates of the LP and FP movies. This is typically accomplished by taking an image of fluorescent beads, such as Invitrogen's FlouSpheres (catalogue no. F8803) that emit fluorescence that is detectable in both the LP and CP channels, and then determining the coordinates of the detected spots, in both channels, through a spatial mapping algorithm. This allows for the matching of individual spots that have both an LP and FP binding. This functionality comes standard in the aforementioned trace creation software packages.

3.5 Trace Idealization

Once the fluorescence trajectories from each single molecule position have been generated, they will need to be idealized. Trace idealization is the process where the raw trace data is converted into a binary signal that identifies when the signal is higher than the background, which corresponds to the FP being bound (Fig. 2). There are many ways of doing this; please *see* **Note 10**.

3.6 Calculating the Fano Factor

A first-pass analysis approach, to determine if the signal indicates changes in the underlying conformation of the riboswitch, is with a Fano factor analysis [22], whereby the Fano factor, a statistical construct defined as the variance divided by the mean of the spikes (FP binding events) is calculated. The behavior of FPs binding a riboswitch, in a single conformation, is accurately modeled with a Poisson distribution [12], and the Fano factor of such a Poisson process is 1 (i.e., the variance and mean are equivalent) [23]. However, if the riboswitch alternates between two conformations that have distinct FP binding kinetics, the signal will not be Poissonian. By calculating the Fano factor, dividing the variance of spike counts by the mean spike count for a given time interval, we can determine if the binding is a simple Poisson process or not. As long as the timescale of riboswitch isomerization is shorter than the entire SiM-KARTS movie taken, then the calculated Fano factor will deviate from 1. It is good practice to calculate Fano factors for various time windows, which gives a quick indication of the timescale of riboswitch isomerization.

3.7 Calculating the Burst and Dwell Times

If the Fano factor analysis indicates more than one riboswitch conformation, then further analysis is warranted to extract burst and nonburst segments of each trace, as described in the following.

1. Feed the idealization of all traces into the burst detection script (https://github.com/RobbWelty/SiMKARTS). The script will produce the burst length time, the dwell time of the interval between spikes (inter-spike interval, or ISI) during and outside of the bursts (i.e., in the nonburst segments) (Fig. 4).

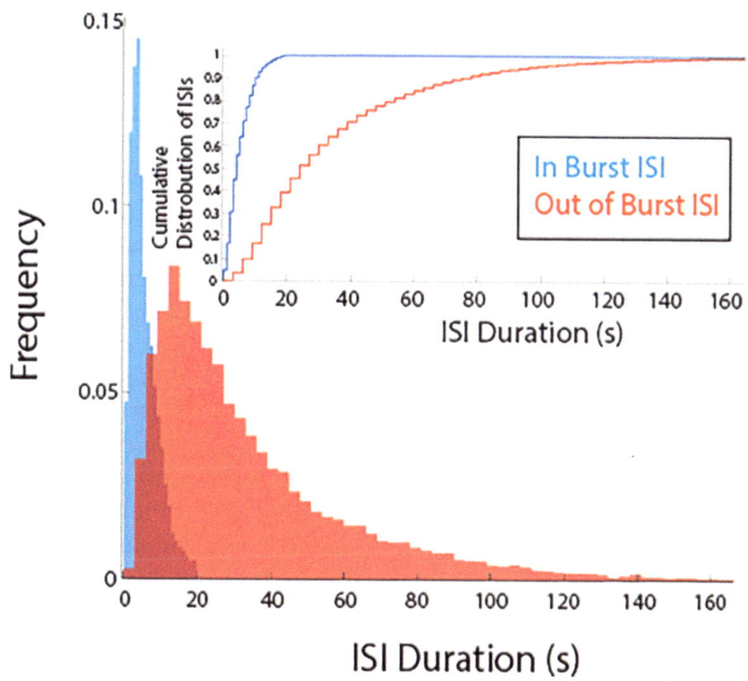

Fig. 4 Normalized histogram representing the distribution of ISIs during bursts (when the binding site is accessible; in blue) and nonburst (when the binding site is not readily accessible; in red). Inset: Cumulative histogram representation of the same data

2. Construct cumulative density plots with the output from the burst detection algorithm.
3. Fit the burst time cumulative distribution function with a single-exponential equation to get the isomerization rate of the riboswitch from SD-accessible to SD-inaccessible conformations.
4. Fit the "in burst" dwell times with a single-exponential function to obtain the binding rate constant of the LP to the more SD-accessible conformation.
5. Fit the "Out of burst", or nonburst, dwell times with a single-exponential function to obtain the binding rate constant of the FP to the more SD-inaccessible conformation.

4 Conclusions

SiM-KARTS represents a powerful method for quantifying RNA structural dynamics on the single-molecule level. By monitoring the binding and dissociation of fluorescently labeled oligonucleotide probes designed to target a specific section(s) of the RNA sequence, the relative populations and the kinetics of isomerization can be inferred via spike train burst analysis.

This technique was primarily designed to interrogate a site-specific structural element of an RNA. Here, we have described an exemplary protocol for the investigation of a preQ1-sensing riboswitch, but SiM-KARTS can be applied to probe any length and type of RNA in its native state, including those directly isolated from a physiological source (e.g., cellular extracts or biofluids) without the need for labeling, which is a significant source of challenge for other single molecule techniques. Due to the rapid exchange of FPs, the observation window can be significantly longer than direct labeling techniques that suffer from slow photobleaching, making it possible to interrogate RNA molecules on the tens-of-minutes timescale. Moreover, SiM-KARTS has the potential to read out secondary as well as tertiary structural dynamics in real time and is mostly limited by the photon emission rates needed to generate a sufficient signal-to-noise ratio at a desired detection speed. In the future, labeling the FP with multiple fluorophores can be used to improve signal-to-noise; and, in principle, any fluorophore-labeled entity (such as a 30S ribosomal subunit) can be used for probing structural change within a biological context of interest [24]. In general, the conceptual design of SiM-KARTS can be adapted to many types of biomolecular interactions in which equilibrium or nonequilibrium processes need to be monitored including not only riboswitches but also, e.g., RNA thermosensors.

5 Notes

1. The OSST composition should take into consideration the specific properties of the biochemical system being targeted. The PCA/PCD system has been observed to affect the rates of enzymes that use NTPs as cofactors [25]. Alternatively, 44 mM glucose, 165 U/mL glucose oxidase from *Aspergillus niger*, in conjunction with 2170 U/mL catalase from *Corynebacterium glutamicum*, can be used for scavenging oxygen [26].
2. To achieve specific immobilization of the target RNA to the microscope slide, the CP (usually DNA, sometimes containing LNA residues for stabilization) should be terminally labeled with biotin. Alternatively, the target RNA can be directly functionalized with biotin, which removes the need for a CP. While direct labeling may seem more efficient, the advantage to using a CP is twofold. First, the same CP potentially can be used with multiple RNA systems. Second, as the RNA of interest is typically the limiting reagent, the labeling efficiency can severely reduce the yield of labeled RNA and introduce additional cumbersome steps that may modify the RNA fold in unexpected ways.
3. LP design is similar to CP but labeled with a fluorophore (usually a Cy3 or equivalent) instead of biotin. Alternatively, instead of using separate CP and LP, a single oligonucleotide can be synthesized bearing both modifications (biotin and fluorophore), allowing it to act as both CP and LP. However, in this case, many more Cy3-labeled spots may be observed that are not associated with an RNA, and no readout of the integrity of a bound RNA is obtained.
4. The key to successful CP and LP design is to ensure that they are bound stably to the RNA transcript throughout the entire experiment. While there are tools for predicting hybridization energies [27] a simple method is to use the web-based melting prediction software provided by commercial synthetic oligonucleotide providers (such as Oligoanalyzer from Integrated DNA Technologies Inc. IDT; or tools from Qiagen). The probes should have a high predicted melting temperature (Tm) of >60 °C (as a proxy for a binding equilibrium constant) and a very low self-structure score (below the experimental temperature). To assure the full integrity of immobilized target RNA, CP and LP should be designed to bind different sections of the RNA sequence, ideally at opposite ends of the target RNA. Modified and HPLC-purified CP and LP can be purchased from commercial suppliers.

5. It is important to verify that the probes do not perturb the folding of the target RNA into its natural secondary structure. If one cannot successfully find sections of the natural RNA sequence to attach probes without perturbing the conformation of interest, one solution is to append arbitrary hybridization sequences to the 5′ and/or 3′ ends of the target RNA, which can hybridize to either CP or LP or both. The sequence of the appended regions can be designed by computational folding predictions, such as Mfold [28] or RNA Structure [29], to avoid interfering with the native fold of the target RNA transcript. The functionality of the target RNA should be checked after hybridization. Efficient hybridization can be verified by nondenaturing gel electrophoresis [30], and the folded secondary structure can be determined via footprinting [31].
6. The FP should have binding and dissociation rates that are (at least) an order of magnitude faster than the isomerization rate of the RNA transcript. SiM-KARTS relies on the rate of the FP binding to the RNA transcript to differ depending on the conformation of the RNA. To optimize for fast FP binding and dissociation, the FP sequence ought to be designed to have a melting temperature close to the experimental temperature. The mean bound time of the FP is ideally 2–5 times longer than the integration time (usually 50–300 ms) of the camera. This helps the trace idealization process and prevents interpreting a single "hot" frame, an artifact caused by electrical fluctuations in the camera, as a binding event. The binding/dissociation rate constants can be empirically optimized by varying the length and GC content of the FP. The dissociation rate constant of a short DNA oligonucleotide, under standard temperature and salt conditions, is exponentially dependent on length. The optimal length of the FP probe with a 50% GC content is between 7 and 8 nucleotides. Modulation of association/dissociation rate constants can be achieved by adjusting the salt concentration (higher monovalent salt concentrations promote binding).
7. The reconstitution strategy will depend on the target system. Systematic optimization (e.g., the annealing temperature, probe concentration) is necessary to achieve a high reconstitution efficiency. A good starting point is to denature the RNA by heating it to 90 °C, include either the CP or LP (as long as they will not bind to an off-target part of the RNA sequence) for 5 min. Let the RNA cool on the benchtop or on ice. It may be helpful to keep the RNA and probes at a high (micromolar) concentration during heating/cooling to maximize hybridization. If this is successful, follow an 8:4:1 annealing strategy

where the LP is eightfold, and the RNA is fourfold, higher in concentration than the CP, which maximizes the probability that the surface immobilized RNA will have both the LP and CP bound. If there is a possibility that the LP or CP will bind to off-target sections of the RNA sequence, then add them after the RNA has cooled.

8. To ensure that the FP is binding to the desired nucleotides of the mRNA sequence, is it beneficial to use a site-blocking oligonucleotide as a further control. This is an RNA or DNA complementary to the FP binding site as well as regions upstream and downstream. If the mRNA is incubated with the site-blocker and prepared for a SiM-KARTS experiment, the FP should be unable to bind the desired binding site. If FP binding events are still observed in the presence of the site-blocking oligonucleotide, it is an indication that the FP is binding also to another part of the mRNA sequence and the design of the FP needs to be reevaluated.
9. Finding the right balance of hardware settings for single molecule experiments can be a challenge, and SiM-KARTS experiments are no exception. The relationship between laser power, probe bound time and camera exposure time needs to be carefully balanced. First, the dwell time of the FP needs to be characterized. Once that time is known, the exposure time of the camera should be adjusted until the mean FP bound time is no less than 3–5 frames, but otherwise as short as possible. This increases temporal resolution while preventing single-frame noise spikes (hot frames) from being counted, which would reduce the accuracy of the data. The laser power should be as high as needed to easily differentiate from bound FP and the background, since higher laser intensities contribute to excess background in some microscope settings. The length of the movie will depend on the rate of FP binding. We recommend taking long (~9000) frame movies to increase the number of binding events observed.
10. The fluorescence signal from single molecule measurements has associated noise. There are many software packages, such as QuB [32], vbFRET [33], and HAMMY [34], that help discern the true fluorescence intensity, a process known as idealization. While it is common for idealization to aim to discern fluorescence intensities for multiple chemical states, SiM-KARTS analysis only needs a binary signal corresponding to whether an LP is bound or not. The Burst analysis script provided at https://github.com/RobbWelty/SiMKARTS requires traces to be idealized, but will automatically reduce any multistate idealization to a binary trace.

Acknowledgments

This work was supported by NIH grant GM131922 to N.G.W. and the Deutsche Forschungsgemeinschaft (DFG, German Research Foundation) Project-ID 449562930 to A.S.

References

1. Al-Hashimi HM, Walter NG (2008) RNA dynamics: it is about time. Curr Opin Struct Biol 18:321–329
2. Cech TR, Steitz JA (2014) The noncoding RNA revolution-trashing old rules to forge new ones. Cell 157:77–94
3. Strobel EJ, Yu AM, Lucks JB (2018) High-throughput determination of RNA structures. Nat Rev Genet 19:615–634
4. Garst AD, Edwards AL, Batey RT (2011) Riboswitches: structures and mechanisms. Cold Spring Harb Perspect Biol 3:a003533
5. Sherwood AV, Henkin TM (2016) Riboswitch-mediated gene regulation: novel RNA architectures dictate gene expression responses. Annu Rev Microbiol 70:361–374
6. Schmidt A, Gao G, Little SR et al (2020) Following the messenger: recent innovations in live cell single molecule fluorescence imaging. WIREs RNA 11:e1587
7. Ray S, Widom JR, Walter NG (2018) Life under the microscope: single-molecule fluorescence highlights the RNA world. Chem Rev 118:4120–4155
8. Blanco M, Walter NG (2010) Chapter 9 – Analysis of complex single-molecule FRET time trajectories. In: WALTER NG (ed) Methods in enzymology. Academic Press
9. Uhm H, Hohng S (2020) Single-Molecule FRET assay for studying cotranscriptional RNA folding. Methods Mol Biol 2016:271–282
10. Ray S, Chauvier A, Walter NG (2019) Kinetics coming into focus: single-molecule microscopy of riboswitch dynamics. RNA Biol 16:1077–1085
11. Chauvier A, Cabello-Villegas J, Walter NG (2019) Probing RNA structure and interaction dynamics at the single molecule level. Methods 162–163:3–11
12. Rinaldi AJ, Lund PE, Blanco MR et al (2016) The Shine-Dalgarno sequence of riboswitch-regulated single mRNAs shows ligand-dependent accessibility bursts. Nat Commun 7:8976
13. Suddala KC, Price IR, Dandpat SS et al (2019) Local-to-global signal transduction at the core of a Mn(2+) sensing riboswitch. Nat Commun 10:4304
14. Suddala KC, Rinaldi AJ, Feng J et al (2013) Single transcriptional and translational preQ1 riboswitches adopt similar pre-folded ensembles that follow distinct folding pathways into the same ligand-bound structure. Nucleic Acids Res 41:10462–10475
15. Eichhorn CD, Kang M, Feigon J (2014) Structure and function of preQ(1) riboswitches. Biochim Biophys Acta 1839:939–950
16. Akimoto H, Imamiya E, Hitaka T et al (1988) Synthesis of queuine, the base of naturally occurring hypermodified nucleoside (queuosine), and its analogues. J Chem Soc Perkin Trans 1:1637–1644
17. Rinaldi AJ, Suddala KC, Walter NG (2015) Native purification and labeling of RNA for single molecule fluorescence studies. Methods Mol Biol 1240:63–95
18. Roy R, Hohng S, Ha T (2008) A practical guide to single-molecule FRET. Nat Methods 5:507–516
19. Gibbs DR, Kaur A, Megalathan A et al (2018) Build your own microscope: step-by-step guide for building a prism-based TIRF microscope. Methods Protoc 1:40
20. Juette MF, Terry DS, Wasserman MR et al (2016) Single-molecule imaging of non-equilibrium molecular ensembles on the millisecond timescale. Nat Methods 13:341–344
21. Hadzic MCA, Kowerko, D, Börner R et al (2016) Detailed analysis of complex single molecule FRET data with the software MASH, SPIE
22. Fano U (1947) Ionization yield of radiations. II. The fluctuations of the number of ions. Phys Rev 72:26–29
23. Eden UT, Kramer MA (2010) Drawing inferences from Fano factor calculations. J Neurosci Methods 190:149–152
24. Chatterjee S, Chauvier A, Dandpat SS et al (2021) A translational riboswitch coordinates nascent transcription–translation coupling. Proc Natl Acad Sci 118:e2023426118

25. Widom JR, Nedialkov YA, Rai V et al (2018) Ligand modulates cross-coupling between riboswitch folding and transcriptional pausing. Mol Cell 72:541–552.e546
26. Aitken CE, Marshall RA, Puglisi JD (2008) An oxygen scavenging system for improvement of dye stability in single-molecule fluorescence experiments. Biophys J 94:1826–1835
27. Markham NR, Zuker M (2005) DINAMelt web server for nucleic acid melting prediction. Nucleic Acids Res 33:W577–W581
28. Zuker M (2003) Mfold web server for nucleic acid folding and hybridization prediction. Nucleic Acids Res 31:3406–3415
29. Reuter JS, Mathews DH (2010) RNA structure: software for RNA secondary structure prediction and analysis. BMC Bioinformatics 11:129
30. Bak G, Han K, Kim KS et al (2015) Electrophoretic mobility shift assay of RNA-RNA complexes. Methods Mol Biol 1240:153–163
31. Wilkinson KA, Merino EJ, Weeks KM (2006) Selective 2′-hydroxyl acylation analyzed by primer extension (SHAPE): quantitative RNA structure analysis at single nucleotide resolution. Nat Protoc 1:1610–1616
32. Qin F, Li L (2004) Model-based fitting of single-channel dwell-time distributions. Biophys J 87:1657–1671
33. Bronson JE, Fei J, Hofman JM et al (2009) Learning rates and states from biophysical time series: a Bayesian approach to model selection and single-molecule FRET data. Biophys J 97: 3196–3205
34. Mckinney SA, Joo C, Ha T (2006) Analysis of single-molecule FRET trajectories using hidden markov modeling. Biophys J 91: 1941–1951

Chapter 5

Isothermal Titration Calorimetry Analysis of a Cooperative Riboswitch Using an Interdependent-Sites Binding Model

Chapin E. Cavender, Griffin M. Schroeder, David H. Mathews, and Joseph E. Wedekind

Abstract

Isothermal titration calorimetry (ITC) is a powerful biophysical tool to characterize energetic profiles of biomacromolecular interactions without any alteration of the underlying chemical structures. In this protocol, we describe procedures for performing, analyzing, and interpreting ITC data obtained from a cooperative riboswitch–ligand interaction.

Key words Thermodynamics, Stoichiometry, Riboswitch, Ligand binding, Parabolic thermogram, Cooperativity, Nonequivalent-interdependent-sites model

1 Introduction

Isothermal titration calorimetry (ITC) is a gold-standard method to determine the thermodynamic interaction profile of biomacromolecular receptors and their ligands. This solution technique directly measures the enthalpic change of a binding reaction, ΔH°, and can fit an apparent equilibrium association constant, K_A, that can be used to calculate the free energy change and the entropy change, ΔG° and ΔS°, as well as the equilibrium dissociation constant K_D (*i.e.*, $1/K_A$). The ratio of bound ligand-to-receptor is also obtained, providing information about the stoichiometry of the interacting complex. Most impressively, ITC provides this information without labels such as fluorescent tags or biotin groups, which can alter the solution behavior of the interacting species, albeit with higher input concentrations of receptor and ligand.

Chapin E. Cavender and Griffin M. Schroeder contributed equally to this work.

Jienyu Ding et al. (eds.), *RNA Structure and Dynamics*, Methods in Molecular Biology, vol. 2568,
https://doi.org/10.1007/978-1-0716-2687-0_5,

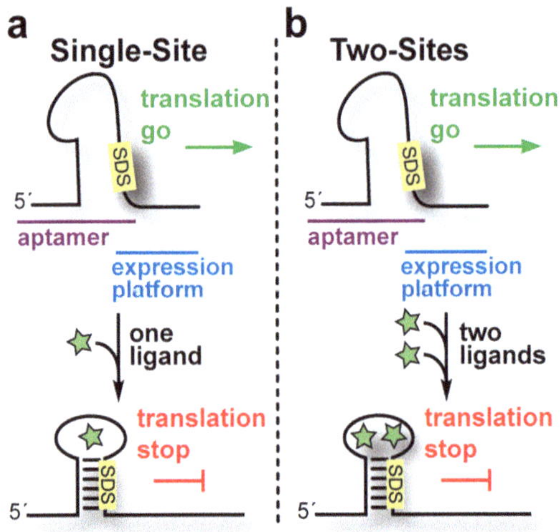

Fig. 1 Schematic drawing of riboswitch organization and gene regulation. (**a**) Riboswitches usually comprise a 5′-aptamer domain that recognizes a ligand with high specificity, and an expression platform that possesses sequences that control gene expression. The example shown indicates the Shine-Dalgarno sequence (SDS) that is exposed during the gene-on state but becomes sequestered upon ligand binding (star), leading to a gene-off conformation. (**b**) An example of two ligands binding in a single aptamer to promote the gene off state. The interplay of the ligands with the receptor leads to positive cooperativity

ITC has been employed extensively to investigate the binding reactions between riboswitches and their cognate cellular ligand (effector) [1–7]. These nonprotein coding (nc)RNAs reside primarily in the 5′ leader sequences of bacterial messenger transcripts. In this context, riboswitches regulate expression of downstream genes by sensing an effector, which promotes conformational changes that control transcription termination or translation initiation (Fig. 1a) [8–10]. Accordingly, effector binding by a riboswitch is integral to gene-regulatory function, since the interplay between riboswitch and effector constitutes a feedback loop that is essential for bacterial metabolic homeostasis, while simultaneously providing a potentially novel target for antimicrobial therapy [11].

Importantly, many structural reports of riboswitches have been enhanced by use of ITC. This approach complements structure determination by measuring the affinity of compounds that are chemically similar to the cognate metabolite, which allows probing the role of individual bases within a binding pocket [4, 6, 7] or through mutational analysis [3, 12]. In some cases, an ITC analysis has been combined with functional assays to examine the role of specific bases in gene regulation [1, 7]. Most riboswitches bind their cognate effectors at a stoichiometric ratio of 1:1 [9]; these thermograms are well-described by a sigmoid-shaped curve that can be fit rigorously by commercially available software. However, these models struggle to describe riboswitches that bind two effectors (Fig. 1b). Such riboswitches continue to present special challenges that yield only qualitative analysis on binding and the extent of cooperativity [5, 13].

Recently, we reported the co-crystal structure of a novel cooperative riboswitch within the class I type I $preQ_1$ ($preQ_1$-I_I) riboswitch family [14] (Fig. 2a, b). Our structure of a representative $preQ_1$-I_I riboswitch from *Carnobacterium antarcticum* (*Can*) revealed that this regulatory ncRNA binds two $preQ_1$ effectors in a single binding pocket wherein the effectors interact with each other by aromatic ring stacking and a shared hydrogen bond, suggestive of cooperative binding (Fig. 2c). ITC experiments confirmed that the binding stoichiometry was indeed two $preQ_1$ metabolites binding to a single riboswitch, as exemplified by a single-phase, sigmoid-shaped thermogram that is consistent with positive cooperativity (Fig. 3a).

To dissect the cooperative interaction, we subjected the *Can* $preQ_1$-I_I riboswitch and related sequences from diverse genera (*i.e.*, *Neisseria gonorrhoeae* and *Haemophilus influenzae*) to ITC binding analysis at 37 °C. Under this condition, each sample revealed parabolic-shaped binding thermograms indicative of dual binding to $preQ_1$. Mutants of key ligand binding nucleobases at 25 °C produced similar curve shapes, suggesting that binding was impaired but positive cooperativity was retained [14]. These curves proved refractory to analysis with available software using a sequential binding sites model, which asserts that one binding site is always populated before the other site and that binding at one site does not alter the likelihood of binding at the other site (*e.g.*, Fig. 3b, c).

Though recently discovered, the mode of dual-$preQ_1$ recognition appears to be conserved in over 1500 known riboswitches of the class I type I $preQ_1$ riboswitch subclass that was first described over a decade ago [15, 16]. Similar thermograms have been observed previously but had not been fit satisfactorily [5]. We hypothesize that other cooperative riboswitches have yet to be discovered and that these ligand–receptor interactions require a more thoughtful and rigorous treatment of binding than currently available programs.

In this chapter we describe a method to collect binding data for our cooperative $preQ_1$-I_I riboswitches and mutants by ITC. To facilitate the accompanying analysis, we developed a new binding model that accounts for two nonequivalent interdependent ligand binding sites and implemented software to fit ITC data to the model. A key feature of our model is that it allows the

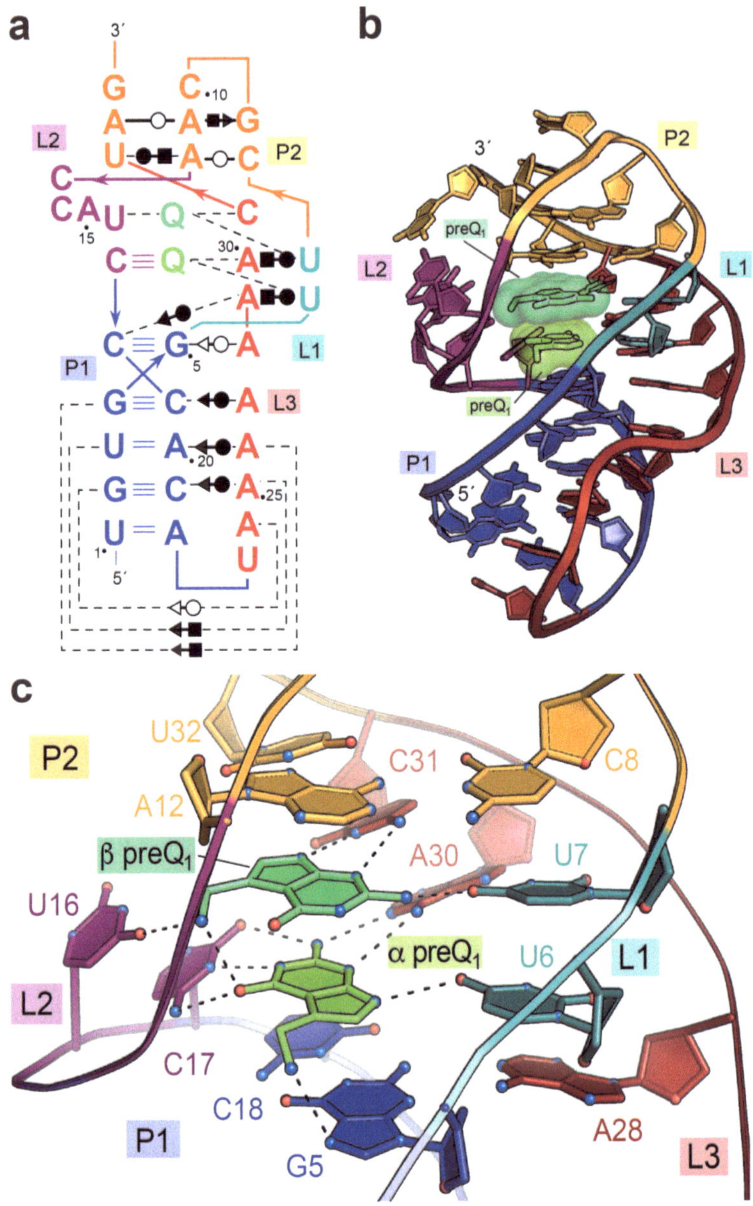

Fig. 2 Secondary structure of the preQ$_1$-I$_I$ riboswitch and ribbon diagrams of the pseudoknot tertiary fold in complex with two preQ$_1$ ligands. (**a**) The *Can* riboswitch with colored sequences corresponding to specific pseudoknot base pairing (P) or loop (L) segments derived from the co-crystal structure [14]. Each preQ$_1$ ligand is labeled "Q." Noncanonical pairing is noted by Leontis–Westhof symbols [36]. (**b**) The *Can* riboswitch fold. The stacked preQ$_1$ ligands are shown as semitransparent surface models. (**c**) Ball-and-stick model of the binding pocket. Dashed lines depict hydrogen bonds to the ligands, which occupy sites labeled "α" and "β." The preQ$_1$ molecules directly contact one another, providing a structural rationale for positive cooperativity

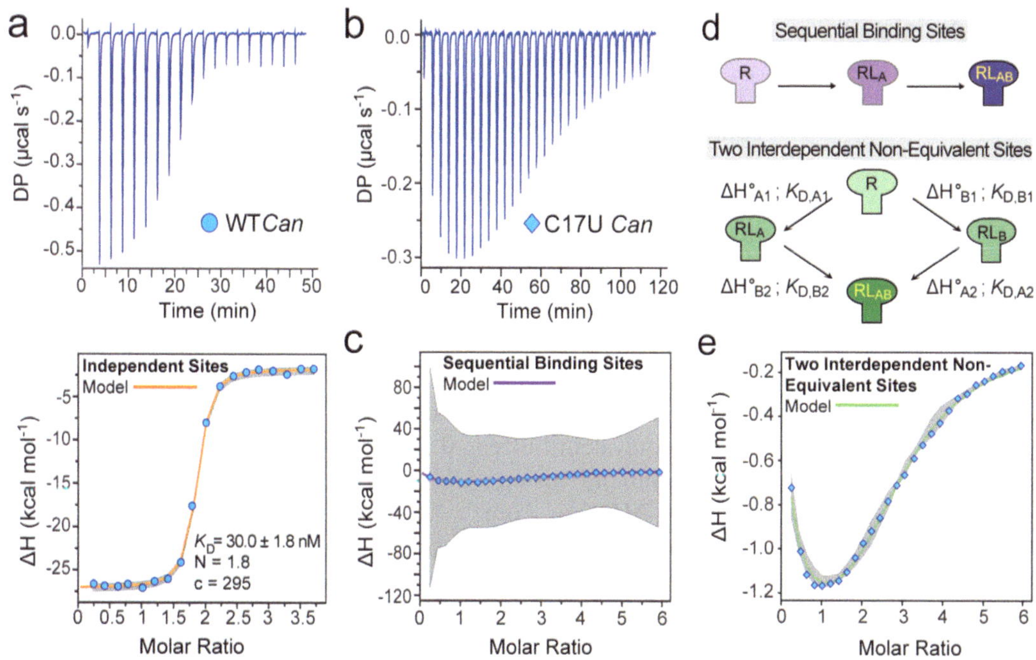

Fig. 3 ITC analysis of the *Can* riboswitch and choice of binding models. (**a**) Representative ITC experiment using the PEAQ ITC (Malvern Panalytical Inc.) with titration of preQ$_1$ into the WT *Can* riboswitch at 25 °C; the binding model is indicated. The resulting apparent equilibrium dissociation constant, K_D, and ligand-to-RNA stoichiometry N are shown; the c value for the experiment is indicated for quality control. Here and elsewhere, the gray zone indicates the 95% confidence intervals from bootstrapping [34]. (**b**) Representative ITC experiment using the VP-ITC (MicroCal) with titration of preQ$_1$ into the C17U *Can* riboswitch at 25 °C. This fit was uninformative as shown by the large uncertainties. (**c**) Use of the sequential binding sites model to fit the titration profile in (**b**). (**d**) Schematic drawings that provide a conceptual framework for the "sequential binding sites" and the "two-interdependent nonequivalent sites" models. (**e**) Use of the "two-interdependent nonequivalent sites" model to fit the titration profile in (**b**). This fit was informative because the uncertainties closely bracket the data points

interdependent ligands to bind the riboswitch in any order (Fig. 3d). Our model provides a sound fit to parabolic thermograms (Fig. 3e) and calculates apparent microscopic equilibrium binding constants, which can be used to derive macroscopic binding constants while providing insight at each step of binding. The ratios of these constants for a particular binding pathway allow quantification of cooperativity of the system by use of defined cooperativity constants. Overall, this method provides a new tool to investigate cooperative binding based on a specific two-interdependent-sites binding model. The Python code is freely available to enable application to comparable problems or the development of derivative approaches through user-based variations to the program.

2 Materials

The following instruments and reagents are needed to successfully conduct an ITC experiment and are applicable to any binding system. It is paramount to avoid ribonuclease contamination during the ITC experiment when analyzing any binding reaction involving an RNA.

2.1 Preparation of Receptor and Ligand

The receptor and ligand must be dissolved in an identical buffer to ensure that the receptor remains folded in a state that is competent for ligand binding throughout the experiment.

1. The ITC buffer comprises: 0.050 M sodium HEPES pH 7.0 (or equivalent buffer), 0.10 M NaCl, and 0.002 M $MgCl_2$. We recommend using UV-irradiated water with a resistivity ≥18 MΩ-cm from a Nanopure UV/UF (ThermoFisher) source or equivalent to prepare this buffer. ITC buffer should be filtered using a 0.2 μm asymmetric Polyethersulfone (aPES) membrane filter (Fisher Scientific) or equivalent and degassed.
   - The reader should note that the ionic concentrations listed above may need to be altered to ensure that the receptor remains folded.
   - If specific metals are needed to facilitate the binding reaction or to fold the receptor, they must be present in this buffer.
2. Folded receptor can be buffer exchanged by dialysis against 2–4 L of ITC buffer overnight at 4 °C using a Slide-A-Lyzer G2 Dialysis Cassette (Thermo-Scientific) or equivalent with a 3500 Da molecular weight cutoff (MWCO).
3. If the reader is purchasing the ligand from a commercial vendor, it should be of the highest possible purity. Solid ligand should be directly dissolved in ITC buffer the day of the experiment to avoid buffer mismatch. A stock of ligand in water prepared at a concentration ≥100-fold excess of that used in the ITC experiment can be diluted in used ITC buffer for the experiment (*i.e.*, the buffer used to dialyze the RNA receptor). This is necessary to avoid a buffer mismatch.
4. The concentrations of receptor and ligand should be determined spectrophotometrically and degassed.

2.2 Instrumentation

The following instruments or equivalents are needed for an ITC experiment: Nanodrop UV Spectrophotometer (Thermo Scientific), pipettes (Gilson), and a calorimeter. We found that an instrument with large cell and syringe volumes, such as a VP-ITC (MicroCal), facilitated this analysis. However, low-volume ITC instruments, such as the PEAQ-ITC (Malvern Panalytical Inc.) or iTC-200 (Malvern Panalytical Inc.), can be used but would require much higher concentrations of sample and ligand to ensure sufficient signal over many injections. The reader will also require access to a computer with an internet connection and Python3 installed.

3 Methods

3.1 Performing the ITC Experiment

ITC requires μg amounts of pure RNA and ligand to load μmol to mmol amounts of sample into the sample cell and syringe, thereby maximizing the signal and to ensure that the results can be replicated two or more times. This chapter provides a practical guide to analyze parabolic thermograms, and we expect that the reader already has a basic understanding of the ITC experiment and the accompanying instrument. Suggested reading includes [17, 18]. A brief description of the experiment is described below.

The receptor, in this case a riboswitch, must be folded without the ligand and subsequently buffer exchanged. Typically, buffer exchanged receptor is loaded into the sample cell of the instrument, whereas the ligand, dissolved in the same exchange buffer, is loaded into the syringe (Fig. 4). The receptor and ligand must be dissolved in the same buffer to avoid undesired heats of dilution resulting from buffer mismatches. Over the course of the experiment, the ligand is titrated into the sample cell and the instrument will measure the amount of power required to keep the sample cell and a reference cell filled with water at the same temperature. Thus, the instrument measures the heats of a binding reaction by recording the power required to maintain a constant temperature between the sample and reference cells at each injection of ligand.

In practice, an ITC instrument has a reliable sensing range and can detect K_D values from 1 nM to 1 mM. The solubility of macromolecule or ligand can impede this [2]. Solubility limitations can complicate reverse titrations that can be useful to investigate cooperative systems. In these experiments, the receptor is loaded into the syringe and titrated into ligand in the sample cell, but this requires a higher concentration of receptor than a usual experiment [19]. Additionally, binding events of interest should be sufficiently fast that the sample cell returns to the reference cell temperature within minutes after each injection, although longer time intervals are possible [20]. We previously described the fundamental steps of an ITC analysis, including RNA folding, as well as step-by-step instructions to perform and analyze a typical experiment [2]. We strongly encourage the reader to consult this protocol for rudimentary 1:1 ligand-to-RNA interaction analysis. Other ITC 1:1 binding examples include high-affinity binding of $preQ_1$ into class II and III $preQ_1$ riboswitches [1, 21, 22], high-affinity binding between RNA–protein complexes [23, 24] and binding between cyclic peptides and RNA [25].

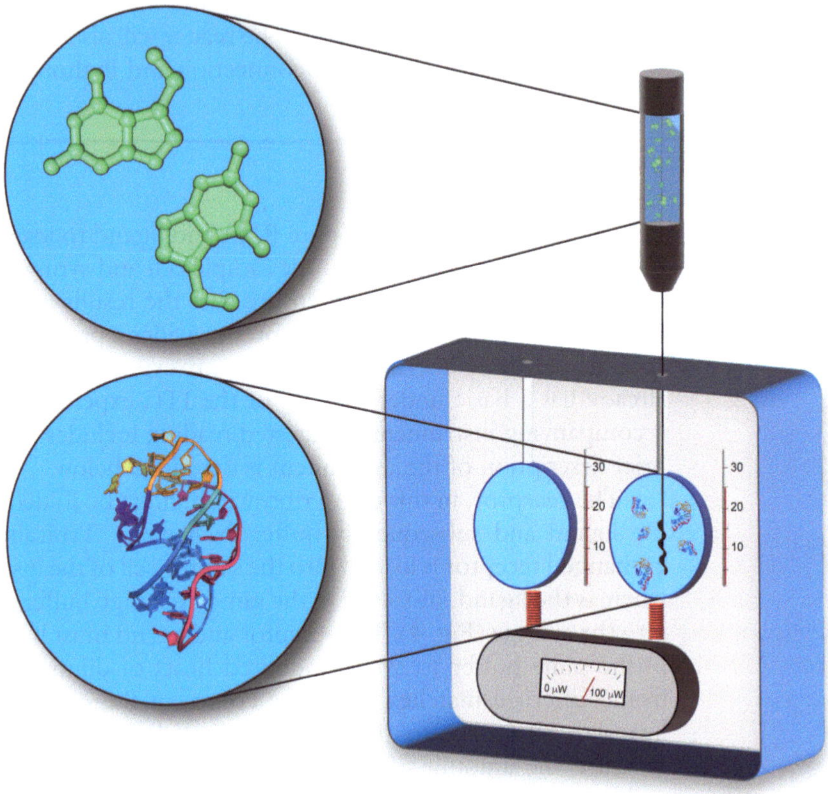

Fig. 4 Schematic diagram of an isothermal titration calorimeter. The calorimeter features two identical cells made of an inert material, such as Hastelloy. The cells are surrounded by an adiabatic jacket. A thermopile or thermocouple is used to detect temperature variations between the reference cell and a sample cell containing the receptor molecule, depicted here as the *Can* $preQ_1$-I_I riboswitch. Before the start of the experiment a feedback circuit is established between the reference and sample cells by application of a small amount of constant power to the reference. This activates a heater (coil) at the sample cell. When the experiment commences, the precisely known quantities of the ligand within the syringe — indicated as the $preQ_1$ metabolite — are injected into the sample cell with mixing at the syringe tip. The resulting heat changes can be exothermic or endothermic. The measured data corresponds to the time-dependent power changes needed to restore the temperature of the sample chamber to that of the reference (indicated by the meter). See [37] for details

3.2 Considerations for Analyzing Parabolic Thermograms

Several important considerations that are applicable to the analysis of parabolic thermogram are the same as those of a typical ITC experiment. The main consideration is that enough sample must be loaded for the instrument to accurately detect the differential power that evolves from the heat of the binding reaction. In our experience, we found that the ratio of 1:20 (riboswitch:ligand) is a reasonable starting point in binding reactions that display parabolic thermograms. However, the concentrations of ligand and riboswitch needed to achieve a sufficient signal will vary depending on the volumes required to load a particular ITC instrument. We found that 3 μM riboswitch and 60 μM ligand were sufficient in most cases using the VP-ITC instrument (MicroCal) over 29 injections with a 240 s spacing and injection volumes of 10 μL. Other binding reactions may not release or absorb as much heat or may require more injections to adequately sample each phase of the thermogram (see below), thus requiring the user to determine the best sample concentrations.

A second consideration is that each phase of the actual binding events should be well populated with data from many injections. For typical ITC experiments, this attribute is described by the unitless *c*-value, in which:

$$c = \frac{[\text{sample in cell}]}{K_D} \times N$$

However, the calculation of a *c*-value does not necessarily hold when examining parabolic curves as the *N*-value (i.e., the number of ligands bound per receptor) is often inaccurate, adding another layer of complexity when analyzing such systems. In practice, one can circumvent this by using multiple injections to populate each phase of the parabolic curve. Although one can easily add more injections to an experiment on any instrument, we have found that an instrument with large cell and syringe volumes, such as the VP-ITC instrument with a 1.8 mL cell volume and ~600 μL syringe volume, provides the sensitivity necessary to perform experiments that are described best by many injections. Additionally, a key feature of our approach (described below) is the ability to globally fit two or more experiments. As such, we encourage the reader to modify the concentrations of ligand between experiments to better sample each phase of the parabola.

3.3 Analysis of Parabolic Thermograms

In this section, we describe the theory behind the interdependent sites binding model (Fig. 5a). Use of this model is justified by structural evidence that shows two $preQ_1$ ligands interacting with one another in their respective binding sites located in a single pocket (Fig. 2c). This finding suggests the need for a cooperative binding model wherein the two effector-binding sites are non-equivalent and interdependent. We implemented this model in a custom in-house Python program based on binding polynomial theory [26].

3.3.1 Theory Behind the Nonequivalent-Interdependent-Sites Binding Model

As noted, fitting an *N*-value during a parabolic thermogram analysis proved unreliable. However, the physical basis of interacting sites is often known from co-crystal structures (*e.g.*, Fig. 2). Therefore, it is reasonable to fix the number of binding sites to exactly two and fit a nuisance variable that represents the effective concentration of active riboswitch in the cell relative to the recorded concentration [26, 27]. Although our binding model describes a binding enthalpy and a microscopic dissociation constant for each

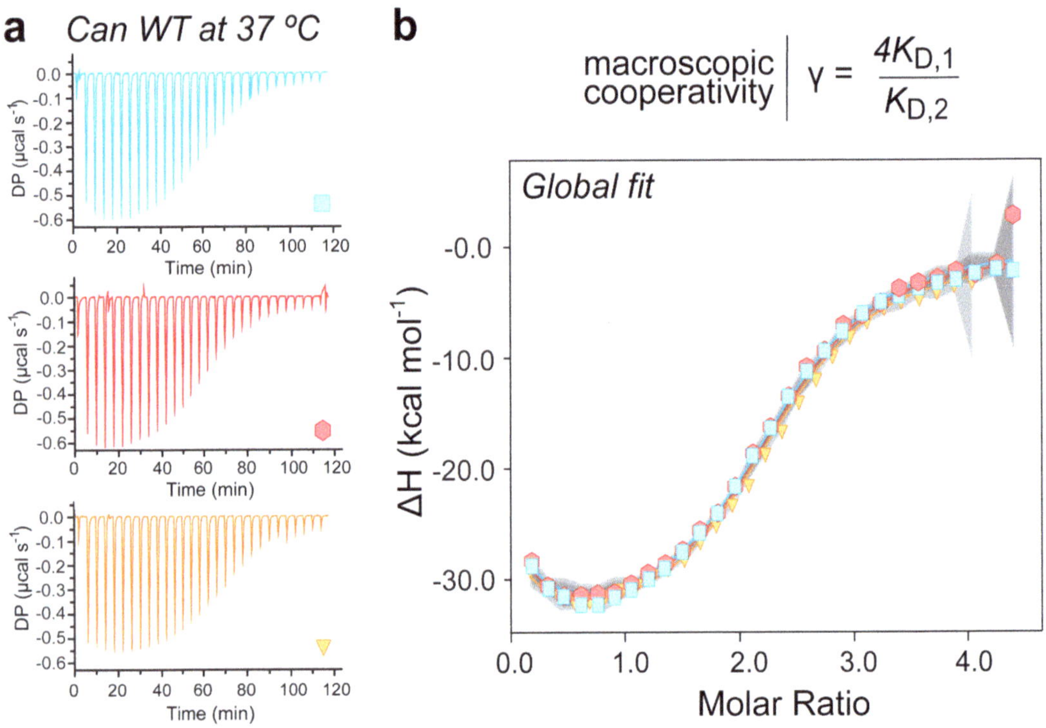

Fig. 5 Global fitting using nonequivalent interdependent sites model. (**a**) Titration of $preQ_1$ into WT *Can* riboswitch at 37 °C performed in triplicate on a VP ITC instrument (MicroCal). (**b**) Global fit of thermograms shown in (**a**). Microscopic binding constants (Fig. 3d) can be used to derive macroscopic binding constants and macroscopic cooperativity

of four distinct binding equilibria (Fig. 3d), there are only three independent microscopic dissociation constants:

$$K_{D,A1} = \frac{[R][L]}{[RL_A]} \quad K_{D,B1} = \frac{[R][L]}{[RL_B]} \quad K_{D,A2} = \frac{[RL_B][L]}{[RL_{AB}]} \quad K_{D,B2} = \frac{[RL_A][L]}{[RL_{AB}]} \tag{1}$$

$$K_{D,A1}K_{D,B2} = K_{D,B1}K_{D,A2} = \frac{[R][L]^2}{[RL_{AB}]} \tag{2}$$

Likewise, there are only three independent binding enthalpies because enthalpy is a state function; completing a thermodynamic cycle must result in no enthalpy change (Fig. 5a).

$$\Delta H^\circ_{A1} + \Delta H^\circ_{B2} - \Delta H^\circ_{A2} - \Delta H^\circ_{B1} = 0 \tag{3}$$

The binding polynomial results in a cubic equation in the concentration of free ligand $[L]$ such that

$$\begin{aligned}&[L]^3 + (2R_T - L_T + K_{D,A2} + K_{D,B2})[L]^2 \\ &+((R_T - L_T)(K_{D,A2} + K_{D,B2}) + K_{D,A1}K_{D,B2})[L] - L_T K_{D,A1}K_{D,B2} = 0\end{aligned} \tag{4}$$

where $R_T = [R] + [RL_A] + [RL_B] + [RL_{AB}]$ is the total concentration of RNA in the ITC cell and $L_T = [L] + [RL_A] + [RL_B] + 2$

$[RL_{AB}]$ is the total concentration of preQ_1 in the ITC cell. We solved this cubic equation analytically by choosing the root that satisfies $[L] = 0$ when $L_T = 0$.

Inspired by recent approaches [28–30], we accounted for the dilution of all chemical species present due to displacement of liquid in the ITC cell by each volume of injected receptor. The differential changes in the concentrations of bound species due to the differential injected volume dV are described by:

$$\begin{aligned} d[RL_A] &= \frac{1}{V_0}(-[RL_A]dV + d\Phi_{A1} - d\Phi_{B2}) \\ d[RL_B] &= \frac{1}{V_0}(-[RL_B]dV + d\Phi_{B1} - d\Phi_{A2}) \\ d[RL_{AB}] &= \frac{1}{V_0}(-[RL_{AB}]dV + d\Phi_{A2} + d\Phi_{B2}) \end{aligned} \tag{5}$$

where V_0 is the volume of the ITC cell and Φ_i is the flux through the binding equilibrium i. The enthalpy can be expressed as a function of the total injected volume V:

$$H(V) = \frac{1}{V_0}\left(\Delta H^\circ_{A1}\frac{d\Phi_{A1}}{dV} + \Delta H^\circ_{B1}\frac{d\Phi_{B1}}{dV} + \Delta H^\circ_{A2}\frac{d\Phi_{A2}}{dV} + \Delta H^\circ_{B2}\frac{d\Phi_{B2}}{dV}\right) \tag{6}$$

The enthalpy change associated with a particular injection that brings the stoichiometric ratio of ligand to receptor S from S_{i-1} to S_i is given by the average value of the enthalpy over this interval:

$$\Delta H_i = \frac{1}{S_i - S_{i-1}}\int_{S_{i-1}}^{S_i} dSH(S) \tag{7}$$

Inserting Eqs. 1, 2, and 3 and Eqs. 5 and 6 into Eq. 7 and using integration by parts gives the injection enthalpy change in terms of the fit parameters, the ITC cell volume, the initial concentrations of RNA in the ITC cell R_0 and of ligand in the syringe L_0, and the concentration of free ligand obtained as the solution to Eq. 4.

$$\begin{aligned} \Delta H_i &= \frac{\Omega(S_i) - \Omega(S_{i-1})}{L_0(S_i - S_{i-1})} \\ \Omega(S) &= \frac{\left(\frac{L_0}{R_0} + S\right)(L_T - [L])(\Delta H_{A1}K_{D,B2} + \Delta H_{B1}K_{D,A2} + (\Delta H_{A1} + \Delta H_{B2})[L])}{K_{D,A2} + K_{D,B2} + 2[L]} \end{aligned} \tag{8}$$

3.3.2 Optimization of Model Parameters

We made use of a trust region reflective algorithm [31] implemented in the optimize.least_squares() method of SciPy [32] to minimize a cost function with the following form:

$$F(\vec{\theta}, \lambda) = \sum_{i=1}^{N}\left(\Delta H_i(\vec{\theta}) - \Delta H_{i,obs}\right)^2 + \lambda\sum_{j=1}^{M}\left(\frac{\theta_j - \theta_{j,0}}{w_j}\right)^2 \tag{9}$$

Here, N represents the number of observed injections and M is the number of fit parameters. The first term is a least-squares minimization that describes the goodness-of-fit between the estimated and observed injection enthalpy changes. The second term is an L2 regularization term, with relative strength controlled by the hyperparameter λ, which prevents overfitting by penalizing deviations of the fit parameters θ from a target value θ_0 (*see* **Note 1**). In a Bayesian framework, this penalty is interpreted as a Gaussian prior on the fit parameters with a mean of θ_0 and a standard deviation of w [33]. For the three independent microscopic dissociation constants, regularization was applied to the natural logarithm of the equilibrium dissociation constant. The regularization targets were set to the values of the fit parameters from a binding model assuming two independent and equivalent binding sites (*i.e.*, the model used by most commercial ITC software). We discuss procedures for choosing the values of the regularization weights and the hyperparameter controlling the regularization strength below. We derived analytical derivatives of the cost function given by Eq. 9 with respect to the fit parameters to take advantage of computationally efficient gradient-based optimization methods.

Selection of Regularization Weights

As a practical guide for selecting regularization weights, the reader is advised to choose values that represent the expected variance in the fit parameter. The regularization targets for the nonequivalent, interdependent binding sites model are derived from the fit parameters using the independent, or single, binding sites model. Since the regularization weight also serves as the standard for a Gaussian prior on the parameter value, the reader can also consider the regularization weight to be a threshold in which the reader would be mildly surprised if the observed value differs from the regularization target by that margin and moderately surprised if the observed value differs by twice that margin.

For most experiments, the regularization weights for the binding enthalpies are $16k_BT$ (roughly 10 kcal mol^{-1} at 37 °C) where k_B is the Boltzmann constant and T is the absolute temperature. We would be mildly surprised if the microscopic binding enthalpies in the nonequivalent, interdependent-binding-sites model differed from the binding enthalpy in the independent-binding-sites model by more than 10 kcal mol^{-1} and moderately surprised if the microscopic binding enthalpies differed by more than 20 kcal mol^{-1}. Similarly, the regularization weight for the nuisance parameter was set to 0.05 because we would be mildly surprised if the effective concentration of binding-competent riboswitch RNA was more than 5% different from the measured concentration and moderately surprised if the effective concentration was more than 10% different.

Albeit these suggestions are general guidelines that can and should be altered in favor of observed patterns in measured data. Particularly, regularization weights that are too small will prevent the optimization from finding a suitable minimum and will result in fits with high bias, or large values of the cost function. Regularization weights that are too large will fail to appropriately regularize the parameter values and will result in fits with high variance and result in extreme values of fit parameters and poor transferability to data not used to train the model parameters. The latter case is diagnosed well by uncertainties obtained from resampling methods such as the bootstrapping procedure described below. For example, the regularization weights of 16 k_BT for the binding enthalpies and $\log_{10}k_BT$ for the logarithms of the dissociation constants produced fits with large, bootstrapped uncertainties for the ITC data from the *Can* C17U sample (Fig. 3b). Thus, we determined that these regularization weights were too large for these sequences and used smaller regularization weights of k_BT for both fit parameters.

Treatment of Uncertainty

To derive estimates and 95% confidence intervals for the fit parameters, we used a bootstrapping method to resample the fitting target in the nonlinear regression [34]. In each bootstrap iteration, we added the residual from the initial fit multiplied by a random number sampled from a standard normal distribution to the observed injection enthalpy changes. The resulting distributions of fit parameters are non-normal, and so we report the estimate of each fit parameter as the median of the bootstrap parameter distribution. We also report a 95% confidence interval as the (2.5, 97.5) percentiles of this bootstrap distribution. The bootstrap was repeated 10,000 times to obtain good estimates of the distribution (*see* **Note 2**). This confidence interval is provided as output and can be plotted as shown in Figs. 3e and 5b.

3.3.3 Global Fitting Provides Better Coverage of Binding Events

A powerful aspect of our approach is the ability to perform a global fit using multiple ITC experiments to obtain a single set of fit parameters (Fig. 5a, b). This allows one to vary the concentration of receptor or ligand to sample different regions of the thermogram, thus providing a composite that yields full coverage of relevant binding events. One offset parameter – a constant added to the estimated injection enthalpy changes – was fit for each experiment. The hyperparameter λ that controls the relative strength of the regularization term was optimized for each riboswitch sequence individually by cross validation across experiments. Each experiment was fit individually for a sequence of λ with logarithmic spacing; specifically, $\log_{10}\lambda$ was varied from -6 to $+6$ in steps of 0.125. The resulting fit parameters were then used to estimate the value of the cost function for the other experiments involving the same riboswitch sequence. The value of λ with the smallest average value of the cost function for experiments not used to train the parameters was chosen for the global fit.

3.3.4 Calculation of Cooperativity

In addition to the fit parameters, our programs will also calculate the following derived parameters: cooperativity C, macroscopic dissociation constants $K_{D,1}$ and $K_{D,2}$, and a macroscopic cooperativity constant called γ (Fig. 5b). These variables are derived from the following relationships:

$$\begin{aligned} C &= \frac{K_{D,A1}}{K_{D,A2}} \\ K_{D,1} &= \left(\frac{1}{K_{D,A1}} + \frac{1}{K_{D,B1}}\right)^{-1} \\ K_{D,2} &= K_{D,A2} + K_{D,B2} \\ \gamma &= \frac{4K_{D,1}}{K_{D,2}} \end{aligned} \tag{10}$$

We obtained estimates and 95% confidence intervals for the derived parameters by calculating the derived parameters for each bootstrap iteration and then reporting the median and (2.5, 97.5) percentile of the bootstrap parameter distribution.

3.4 Application of the Interdependent Sites Model

After performing the ITC experiment, some minor adjustments of the baseline might be appropriate to improve the fit [2] as a preface to the use of our software. Such adjustment can be made using PEAQ-ITC software (Malvern Panalytical Inc.), Origin v7.0 software (MicroCal) or comparable programs. Subsequently, the injection data must be exported as a space delimited file. The output data typically includes the following information at each injection: differential power (ΔQ), volume injected, the concentration of injectant in the cell (X_t), the concentration of cell material in the active volume (M_t), ΔH, the fit, and residuals measured at each injection.

3.4.1 Preparing ITC Data for Analysis

In this section, we describe how to prepare parabolic thermograms and export the relevant data for analysis by our model using software associated with the VP-ITC and PEAQ-ITC instruments. A text file with the fitting parameters can be exported by most analysis software. Thus, we encourage the reader to follow the guide below and the manual for their preferred analysis software to prepare data for our python code.

PEAQ-ITC:

1. Load the .itc experiment file into the PEAQ-ITC analysis software. The software will provide a rudimentary fit using a single site binding model.

2. The reader is encouraged to adjust the baseline and integrations as necessary in the adjusted baseline window as described in the manual and [2] (*see* **Note 3**).
3. Once satisfied that your fit accurately matches the data, navigate to the presentation window in the upper control bar.
4. While in presentation mode, navigate to the injection table and export as a .csv file. This can be opened in another program, such as Microsoft Excel, and exported as a whitespace-delimited file, where the whitespace can be spaces or Tables.

VP-ITC:

1. Load the .itc experiment file into the Origin v7.0 analysis software and adjust the baseline as necessary (stated above). The reader is encouraged to consult the manual and [2] for a guide to the Origin v7.0 software.
2. Open the worksheet by selecting "plot" in the format menu. A dialog box will open for your thermogram, select worksheet.
3. This will bring up a table with the injection parameters. This data can be copied and pasted into another program, such as Microsoft Excel, and exported as a whitespace-delimited file, where the whitespace can be spaces or tabs.

3.4.2 Running the Python Program

Once the file is created, the input to our Python code can be prepared by retaining the columns corresponding to the injected volume in μL and the measured enthalpies in cal mol^{-1} exported as a whitespace-delimited text file. In each row, the measured enthalpy in the second column should be the enthalpy measured after the injection containing the volume in the first column. Some commercial ITC software records the measured enthalpy in the row for the subsequent injection such that the row for the first injection records no measured enthalpy and the last row records an enthalpy with no injected volume. In such cases, the measured enthalpies should be shifted up one row to align with the injected volumes.

1. Now, download the relevant project files from GitHub at: https://github.com/chapincavender/itc_two_site_fit. The python code is fit_itc_model.py. The README (shown on the project front page) explains the input parameters. There are two python dependencies that need to be installed, numpy [35] and scipy [32].
2. Sample input files for the data analyzed in this chapter and [14] are provided at https://github.com/chapincavender/itc_two_site_fit/tree/main/2021_preq1_riboswitch_data/itc_data with an example shown in Fig. 6a.

a

C17U replicate 1 input

Injected Volume (μL)	Measured Enthalpy (kcal mol^{-1})
3.000000000000000000e+00	-2.785470599999999877e+02
1.000000000000000000e+01	-6.957294600000000173e+02
1.000000000000000000e+01	-9.924577799999999570e+02
1.000000000000000000e+01	-1.094715609999999970e+03
1.000000000000000000e+01	-1.144753020000000106e+03
1.000000000000000000e+01	-1.158762500000000045e+03
1.000000000000000000e+01	-1.157542369999999892e+03
1.000000000000000000e+01	-1.121187419999999975e+03
1.000000000000000000e+01	-1.081119860000000017e+03
1.000000000000000000e+01	-1.031627400000000080e+03
1.000000000000000000e+01	-9.676027199999999766e+02
1.000000000000000000e+01	-8.949441399999999476e+02
1.000000000000000000e+01	-8.296408099999999877e+02
1.000000000000000000e+01	-7.574937099999999646e+02
1.000000000000000000e+01	-7.085965599999999540e+02
1.000000000000000000e+01	-6.507267600000000130e+02
1.000000000000000000e+01	-5.742534100000000308e+02
1.000000000000000000e+01	-5.184008599999999660e+02
1.000000000000000000e+01	-4.730613399999999729e+02
1.000000000000000000e+01	-4.250599700000000212e+02
1.000000000000000000e+01	-3.745905200000000264e+02
1.000000000000000000e+01	-3.311435700000000111e+02
1.000000000000000000e+01	-2.980179699999999912e+02
1.000000000000000000e+01	-2.672516100000000279e+02
1.000000000000000000e+01	-2.361283799999999928e+02
1.000000000000000000e+01	-2.165124999999999886e+02
1.000000000000000000e+01	-1.939138800000000060e+02
1.000000000000000000e+01	-1.863420800000000099e+02
1.000000000000000000e+01	-1.735986100000000079e+02

C17U replicate 2 input

Injected Volume (μL)	Measured Enthalpy (kcal mol^{-1})
3.000000000000000000e+00	-2.869007500000000164e+02
1.000000000000000000e+01	-7.250242399999999634e+02
1.000000000000000000e+01	-1.013587620000000015e+03
1.000000000000000000e+01	-1.114517640000000029e+03
1.000000000000000000e+01	-1.161277859999999919e+03
1.000000000000000000e+01	-1.166182780000000093e+03
1.000000000000000000e+01	-1.158595530000000053e+03
1.000000000000000000e+01	-1.141218399999999974e+03
1.000000000000000000e+01	-1.102685919999999896e+03
1.000000000000000000e+01	-1.040134180000000015e+03
1.000000000000000000e+01	-9.747905799999999772e+02
1.000000000000000000e+01	-9.186315299999999979e+02
1.000000000000000000e+01	-8.597031799999999748e+02
1.000000000000000000e+01	-7.871070999999999458e+02
1.000000000000000000e+01	-7.126670800000000554e+02
1.000000000000000000e+01	-6.654017000000000053e+02
1.000000000000000000e+01	-5.896357100000000173e+02
1.000000000000000000e+01	-5.325228799999999865e+02
1.000000000000000000e+01	-4.786514199999999732e+02
1.000000000000000000e+01	-4.301716999999999871e+02
1.000000000000000000e+01	-3.732841500000000110e+02
1.000000000000000000e+01	-3.169383000000000266e+02
1.000000000000000000e+01	-2.959704899999999839e+02
1.000000000000000000e+01	-2.558529200000000117e+02
1.000000000000000000e+01	-2.404666899999999998e+02
1.000000000000000000e+01	-2.138593700000000126e+02
1.000000000000000000e+01	-1.950181100000000072e+02
1.000000000000000000e+01	-1.839962400000000002e+02
1.000000000000000000e+01	-1.654046099999999910e+02

b

C17U global fit output

```
# Experiment 0 Ligand   81.00000000 uM Receptor   3.00000000 uM Volume  1420.60000000 uL
# Experiment 1 Ligand   81.00000000 uM Receptor   3.00000000 uM Volume  1420.60000000 uL
# Guess_cost        0.41829981 (kcal_mol^-1)^2 Guess_chi_sq       0.00890000
# Fit_cost          0.02209137 (kcal_mol^-1)^2 Fit_chi_sq         0.00047003
# Bootstrap_cost    0.02115120 (kcal_mol^-1)^2 Bootstrap_chi_sq   0.00045003
```

Parameter	Fit_value	Confidence_lower	Confidence_upper	Bootstrap_value	Bootstrap_lower	Bootstrap_upper
Offset_1_(kcal_mol^-1)	-0.06438990	-0.10701724	-0.02176255	-0.06279564	-0.07805474	-0.02333700
Offset_2_(kcal_mol^-1)	-0.07360844	-0.11623579	-0.03098110	-0.07190413	-0.08686652	-0.03235193
Eta	0.99695329	0.94400453	1.04990204	0.99118836	0.94371330	1.03100600
DeltaH_A1_(kcal_mol^-1)	-1.65952062	-2.52885938	-0.79018185	-1.63076953	-1.78988574	0.23370898
DeltaH_B1_(kcal_mol^-1)	0.37450353	-0.75618222	1.50518927	0.35330953	-1.88115693	0.82183202
DeltaH_A2_(kcal_mol^-1)]	-4.30304549	-5.96906784	-2.63702315	-4.33627563	-4.89886733	-2.25786509
DeltaH_B2_(kcal_mol^-1)	-2.26902135	-3.13008483	-1.40795787	-2.32931291	-4.38614774	-2.17057247
DeltaG_A1_(kcal_mol^-1)	1.03907553	0.56197907	1.51617200	1.03037804	0.97647245	1.07987922
DeltaG_B1_(kcal_mol^-1)	1.13724656	0.29645981	1.97803331	1.14881903	1.08203440	1.24523880
DeltaG_A2_(kcal_mol^-1)	-0.31877509	-0.83090719	0.19335701	-0.31584081	-0.42014742	-0.12969819
DeltaG_B2_(kcal_mol^-1)	-0.22060406	-0.68646030	0.24525217	-0.19788389	-0.29803939	-0.01133760
KD_A1_(uM)	5.77627302	2.58186512	12.92295626	5.69209844	5.19707913	6.18809573
KD_B1_(uM)	6.81722422	1.64932623	28.17789786	6.95168777	6.21064614	8.18024757
KD_A2_(uM)	0.58389697	0.24600320	1.38589933	0.58679589	0.49207327	0.80339765
KD_B2_(uM)	0.68912196	0.31392151	1.51276374	0.71606112	0.60469391	0.98104630
Alpha	9.89262373	3.03570860	32.23761476	9.72096687	7.31410411	11.21325634
KD_1_(uM)	3.12686362	1.18854530	8.22625448	3.12768093	3.02935765	3.22431014
KD_2_(uM)	1.27301893	0.71156621	2.27747913	1.29995575	1.13819093	1.74598240
Gamma	9.82503418	3.16752249	30.47533110	9.60050268	7.19528503	11.02776344

Stoichiometric_ratio	Target_enthalpy_(kcal_mol^-1)	Fit_enthalpy	Confidence_lower	Confidence_upper	Bootstrap_enthalpy	Bootstrap_lower	Bootstrap_upper
0.24821267	-0.69572946	-0.76259625	-1.16618034	-0.35901216	-0.76030026	-0.85971758	-0.66036050
0.44069713	-0.99245778	-0.96874903	-1.38163438	-0.55586369	-0.96851441	-1.01994702	-0.91681345

Fig. 6 Example input and output for interdependent-sites-binding model code. (**a**) Representative input dataset from ITC experiments of $preQ_1$ titrated into *Can* C17U at 25 °C. The first column is the volume (μL) of $preQ_1$ injected into the cell while the second column is the resulting enthalpies (kcal mol^{-1}). (**b**) Example globally fit output of C17U input data shown in (**a**). The labels A1 and B1 correspond to Fig. 3d. Each parameter fit value is shown along with the bootstrapped value and the 95% confidence intervals

3. Similarly, sample output is provided at https://github.com/chapincavender/itc_two_site_fit/tree/main/2021_preq1_riboswitch_data/fits with an example shown in Fig. 6b.
4. To reproduce the example in Fig. 6, the Python command line is:
5. `python3 fit_itc_model.py -t 298.15 -p 0.00042169650342858224 -n 2 -b 10,000 -l 81,81 -r 3,3 -v 1420.6 -s 1 can_c17u_preq1_1_peaq_data can_c17u_preq1_2_peaq_data`,

6. A description of relevant input parameters is as follows: -t 298.15 indicates the temperature in K (to determine the widths of the priors and to determine free energy changes); −p indicates the L2 regularization term in $kcal^2\ mol^{-2}$ (where the value provided was determined by cross validation and is $10^{-3.375}$); −n 2 indicates two binding sites; −b 10,000 is the number of bootstrap calculations to perform; −l 81,81 indicates that the two experiments had 81 μM of preQ1 in the syringe; −r 3,3 indicates that both experiments had 3 μM of riboswitch in the cell; −v 1420.6 is the μL volume of the cell; −s 1 indicates that the first injection is not included in the fit; and the input files can_c17u_preq1_1_peaq_data and can_-c17u_preq1_2_peaq_data are available in the GitHub project.

3.5 Interpretation of Output Data

Use of our program to fit parabolic titrations from WT, C17U, and C31U *Can* riboswitch sequences (*e.g.*, Fig. 3b) yielded the results in Figs. 3e, 5b, and Table 1 as well as other WT $preQ_1$-I_I sequences from *N. gonorrhoeae* and *H. influenzae* [14]. Each binding event is well sampled by data points using the global fitting approach (Fig. 5b). The error-of-the-fit for the two-interdependent non-identical sites model is superior to that of the sequential binding sites model (gray shading in Fig. 3c, e). Although titrations of WT sequences were recorded at 37 °C to accentuate the cooperative character of $preQ_1$ binding [14], a comparison of the WT *Can* result to those of the C17U and C31U mutants, recorded at 25 °C, indicates substantial impairment of ligand binding based on inspection of microscopic ($K_{D,A1}$, $K_{D,A2}$, $K_{D,B1}$, and $K_{D,B2}$) and macroscopic ($K_{D,1}$ and $K_{D,2}$) K_D values. Despite weakened K_D values and reduced ΔH° values for most microscopic binding steps (Table 1), both the C17U (Fig. 3b) and C31U retained parabolic isotherms, indicating that they retained cooperative binding [14]. Based upon the ratio of microscopic binding constants, a cooperativity constant C is defined, which indicates positive cooperativity when >1.0. Accordingly, WT and mutant sequences show positive cooperativity with C constants of 8.1, 9.7, and 3.0 (Table 1). A similar conclusion can be drawn based upon the macroscopic binding constant γ (Fig. 5b and Table 1). Notably, C31U produced the most detrimental effect on $preQ_1$ binding and cooperativity. A plausible explanation is that C31 supports β-site ligand binding by forming the pocket ceiling. This stacking scaffold establishes a key contact point for β-site $preQ_1$ binding that is necessary for efficient binding within the underlying α position (Fig. 2a–c).

3.6 Code Availability

The ITC fitting software and parameter fits are available on GitHub at https://github.com/chapincavender/itc_two_site_fit, distributed under the MIT license. Overall, we anticipate that our program will be useful for analysis of other cooperative riboswitches. The availability of our program allows easy access while also permitting code modification for derivative studies.

Table 1
Average Thermodynamic Parameters for the *Can* PreQ$_1$-I$_I$ Riboswitch

Parameter	*Can*[a] WT	*Can*[b] C17U	*Can*[b] C31U
$\Delta H°_{A1}$ (kcal mol^{-1})	−36.9 (−37.8, −36.3)[c]	−1.6 (−1.8, 0.2)	8.6 (7.6, 9.4)
$\Delta H°_{B1}$ (kcal mol^{-1})	−29.9 (−31.4, −28.6)	0.4 (−1.9, 0.8)	−17.7 (−18.7, −15.4)
$\Delta H°_{A2}$ (kcal mol^{-1})	−47.4 (−48.8, −45.7)	−4.3 (−4.9, −2.3)	−23.5 (−25.6, −22.4)
$\Delta H°_{B2}$ (kcal mol^{-1})	−40.3 (−40.9, −39.5)	−2.3 (−4.4, −2.2)	−49.7 (−51.3, −48.0)
$K_{D,A1}$ (nM)	1480.0 (1190.0, 1840.0)	5690.0 (5200.0, 6190.0)	9660.0 (8640.0, 11050.0)
$K_{D,B1}$ (nM)	2260.0 (1300.0, 3650.0)	6950.0 (6210.0, 8180.0)	21300.0 (17700.0, 25300.0)
$K_{D,A2}$ (nM)	182.0 (126.0, 273.0)	587.0 (492.0, 803.0)	3200.0 (2430.0, 4180.0)
$K_{D,B2}$ (nM)	278.0 (249.0, 305.0)	716.0 (605.0, 981.0)	7020.0 (6560.0, 7400.0)
C[d]	8.1 (4.5, 14.3)	9.7 (7.3, 11.2)	3.0 (2.6, 3.6)
$K_{D,1}$[e] (nM)	891.0 (630.0, 1208.0)	3130.0 (3030.0, 3220.0)	6640.0 (6420.0, 6830.0)
$K_{D,2}$ (nM)	461.0 (380.0, 565.0)	1300.0 (1140.0, 1750.0)	10260.0 (9460.0, 10980.0)
γ[f]	7.7 (4.5, 12.6)	9.6 (7.2, 11.0)	2.6 (2.5, 2.7)

[a]Sequence from the *Carnobacterium antarcticum* (*Can*) class I type I riboswitch
[b]ITC was recorded at 25 °C for mutants; WT data were recorded at 37 °C
[c]Values are the median and confidence intervals (2.5, 97.5) from bootstrapping
[d]The ratio of microscopic binding constants gives the cooperativity constant C. See the text
[e]The microscopic dissociation constants can be used to generate the macroscopic binding constants, $K_{D,1}$ and $K_{D,2}$, that correspond to the first and second ligand binding steps (Fig. 3d)
[f]The ratio of macroscopic binding constants ($K_{D,1}$, $K_{D,2}$) multiplied by a statistical factor of 4 gives the macroscopic cooperativity constant γ as described in the Eq. 10

4 Notes

1. The L2 regularization term in Eq. 9 is determined via cross validation by scanning multiple values. A sample bash script, fit_2021_preq1_riboswitch_data.sh, is provided as part of the GitHub project. It demonstrates the approach for automating the determination of the term.

2. The reader should note that the bootstraps are stochastic, and therefore the exact numbers will vary starting in approximately the fourth significant figure.
3. The goal of this step is to gently guide the fit to match the shape of the parabolic curve. The reader must avoid excluding large areas of each peak from integration to ensure a rigorous fit. The reader should not be discouraged that the reduced chi-squared value will likely remain high after any adjustments. This value refers to the goodness-of-fit of the single-site model to the parabolic curve, which will never describe a parabolic fit well.

Acknowledgments

This work was supported by NIH grants GM132185 to D.H.M and GM063162 to J.E.W. G.M.S. was supported by training grant GM118283, and C.E.C was supported by training grant AI049815 to J.E.W. and an Elon Huntington Hooker graduate fellowship.

References

1. Dutta D, Wedekind JE (2020) Nucleobase mutants of a bacterial preQ$_1$-II riboswitch that uncouple metabolite sensing from gene regulation. J Biol Chem 295(9):2555–2567
2. Liberman JA, Bogue JT, Jenkins JL, Salim M, Wedekind JE (2014) ITC analysis of ligand binding to preQ$_1$ riboswitches. Methods Enzymol 549:435–450
3. Schroeder GM, Dutta D, Cavender CE, Jenkins JL, Pritchett EM, Baker CD, Ashton JM, Mathews DH, Wedekind JE (2020) Analysis of a preQ$_1$-I riboswitch in effector-free and bound states reveals a metabolite-programmed nucleobase-stacking spine that controls gene regulation. Nucleic Acids Res 48(14): 1846–1864
4. Chen H, Egger M, Xu X, Flemmich L, Krasheninina O, Sun A, Micura R, Ren A (2020) Structural distinctions between NAD+ riboswitch domains 1 and 2 determine differential folding and ligand binding. Nucleic Acids Res 48(21):12394–12406
5. Huang L, Serganov A, Patel DJ (2010) Structural insights into ligand recognition by a sensing domain of the cooperative glycine riboswitch. Mol Cell 40(5):774–786
6. Ren A, Patel DJ (2014) c-di-AMP binds the ydaO riboswitch in two pseudo-symmetry-related pockets. Nat Chem Biol 10(9): 780–786
7. Trausch JJ, Batey RT (2018) A disconnect between high-affinity binding and efficient regulation by antifolates and purines in the tetrahydrofolate riboswitch. Chem Biol 21(2): 205–216
8. Breaker RR (2018) Riboswitches and translation control. Cold Spring Harb Perspect Biol 10:a032797
9. McCown PJ, Corbino KA, Stav S, Sherlock ME, Breaker RR (2017) Riboswitch diversity and distribution. RNA 23(7):995–1011
10. Nahvi A, Sudarsan N, Ebert MS, Zou X, Brown KL, Breaker RR (2002) Genetic control by a metabolite binding mRNA. Chem Biol 9(9):1043–1049
11. Blount KF, Breaker RR (2006) Riboswitches as antibacterial drug targets. Nat Biotechnol 24(12):1558–1564
12. Trausch JJ, Ceres P, Reyes FE, Batey RT (2011) The structure of a tetrahydrofolate-sensing riboswitch reveals two ligand binding sites in a single aptamer. Structure 19(10): 1413–1423
13. Gao A, Serganov A (2014) Structural insights into recognition of c-di-AMP by the ydaO riboswitch. Nat Chem Biol 10(9):787–792
14. Schroeder GM, Cavender CE, Blau ME, Jenkins JL, Mathews DH, Wedekind JE (2022) A small RNA that cooperatively senses two

stacked metabolites in one pocket for gene control. Nat Commun. 13(1): 199

15. McCown PJ, Liang JJ, Weinberg Z, Breaker RR (2014) Structural, functional, and taxonomic diversity of three preQ$_1$ riboswitch classes. Chem Biol 21(7):880–889
16. Roth A, Winkler WC, Regulski EE, Lee BW, Lim J, Jona I, Barrick JE, Ritwik A, Kim JN, Welz R, Iwata-Reuyl D, Breaker RR (2007) A riboswitch selective for the queuosine precursor preQ$_1$ contains an unusually small aptamer domain. Nat Struct Mol Biol 14(4):308–317
17. Velazquez-Campoy A, Leavitt E, Freire SF (2004) Characterization of protein-protein interactions by isothermal titration calorimetry. Methods Mol Biol 261:35–54
18. Freire E, Mayorga OL, Straume M (1990) Isothermal titration calorimetry. Anal Chem 62(18):950–959
19. Brown A (2009) Analysis of cooperativity by isothermal titration calorimetry. Int J Mol Sci 10(8):3457–3477
20. Vander Meulen KA, Butcher SE (2012) Characterization of the kinetic and thermodynamic landscape of RNA folding using a novel application of isothermal titration calorimetry. Nucleic Acids Res 40(5):2140–2151
21. Liberman JA, Suddala KC, Aytenfisu A, Chan D, Belashov IA, Salim M, Mathews DH, Spitale RC, Walter NG, Wedekind JE (2015) Structural analysis of a class III preQ$_1$ riboswitch reveals an aptamer distant from a ribosome-binding site regulated by fast dynamics. Proc Natl Acad Sci USA 112(27): E3485–E3494
22. Liberman JA, Salim M, Krucinska J, Wedekind JE (2013) Structure of a class II preQ$_1$ riboswitch reveals ligand recognition by a new fold. Nat Chem Biol 9(6):353–355
23. Chavali SS, Mali SM, Jenkins JL, Fasan R, Wedekind JE (2020) Co-crystal structures of HIV TAR RNA bound to lab-evolved proteins show key roles for arginine relevant to the design of cyclic peptide TAR inhibitors. J Biol Chem 295(49):16470–16486
24. Belashov IA, Crawford DW, Cavender CE, Dai P, Beardslee PC, Mathews DH, Pentelute BL, McNaughton BR, Wedekind JE (2018) Structure of HIV TAR in complex with a lab-evolved RRM provides insight into duplex RNA recognition and synthesis of a constrained peptide that impairs transcription. Nucleic Acids Res 46(130):6401–6415
25. Chavali SS, Mali SM, Bonn R, Saseendran Anitha A, Bennett RP, Smith HC, Fasan R, Wedekind JE (2021) Cyclic peptides with a distinct arginine-fork motif recognize the HIV trans-activation response RNA *in vitro* and in cells. J Biol Chem 297(6):101390
26. Freire E, Schön A, Velazquez-Campoy A (2009) Chapter 5 isothermal titration calorimetry: general formalism using binding polynomials. Methods Enzymol 455:127–155
27. Duvvuri H, Wheeler LC, Harms MJ (2018) pytc: Open-source python software for global analyses of isothermal titration calorimetry data. Biochemistry 57(18):2578–2583
28. Herrera I, Winnik MA (2016) Differential binding models for direct and reverse isothermal titration calorimetry. J Phys Chem B 120(9):2077–2086
29. Herrera I, Winnik MA (2013) Differential binding models for isothermal titration calorimetry: moving beyond the Wiseman isotherm. J Phys Chem B 117(29):8659–8672
30. Sigurskjold BW (2000) Exact analysis of competition ligand binding by displacement isothermal titration calorimetry. Anal Biochem 277(2):260–266
31. Branch MA, Coleman TF, Li Y (1999) A subspace, interior, and conjugate gradient method for large-scale bound-constrained minimization problems, SIAM. J Sci Comput 21(1): 1–23
32. Virtanen P, Gommers R, Oliphant TE, Haberland M, Reddy T, Cournapeau D, Burovski E, Peterson P, Weckesser W, Bright J, van der Walt SJ, Brett M, Wilson J, Millman KJ, Mayorov N, Nelson ARJ, Jones E, Kern R, Larson E, Carey CJ, Polat İ, Feng Y, Moore EW, VanderPlas J, Laxalde D, Perktold J, Cimrman R, Henriksen I, Quintero EA, Harris CR, Archibald AM, Ribeiro AH, Pedregosa F, van Mulbregt P, Vijaykumar A, Bardelli AP, Rothberg A, Hilboll A, Kloeckner A, Scopatz A, Lee A, Rokem A, Woods CN, Fulton C, Masson C, Häggström C, Fitzgerald C, Nicholson DA, Hagen DR, Pasechnik DV, Olivetti E, Martin E, Wieser E, Silva F, Lenders F, Wilhelm F, Young G, Price GA, Ingold G-L, Allen GE, Lee GR, Audren H, Probst I, Dietrich JP, Silterra J, Webber JT, Slavič J, Nothman J, Buchner J, Kulick J, Schönberger JL, de Miranda Cardoso JV, Reimer J, Harrington J, Rodríguez JLC, Nunez-Iglesias-J, Kuczynski J, Tritz K, Thoma M, Newville M, Kümmerer M, Bolingbroke M, Tartre M, Pak M, Smith NJ, Nowaczyk N, Shebanov N, Pavlyk O, Brodtkorb PA, Lee P, McGibbon RT, Feldbauer R, Lewis S, Tygier S, Sievert S, Vigna S, Peterson S, More S, Pudlik T, Oshima T, Pingel TJ, Robitaille TP, Spura T, Jones TR, Cera T, Leslie T, Zito T, Krauss T, Upadhyay U, Halchenko YO, Vázquez-

Baeza Y, SciPy C (2020) SciPy 1.0: fundamental algorithms for scientific computing in Python. Nat Methods 17(3):261–272

33. Williams PM (1995) Bayesian regularization and pruning using a laplace prior. Neural Comput 7(1):117–143

34. Wu CFJ (1986) Jackknife, bootstrap and other resampling methods in regression analysis. Ann Stat 14(4):1261–1295

35. Harris CR, Millman KJ, van der Walt SJ, Gommers R, Virtanen P, Cournapeau D, Wieser E, Taylor J, Berg S, Smith NJ, Kern R, Picus M, Hoyer S, van Kerkwijk MH, Brett M, Haldane A, del Río JF, Wiebe M, Peterson P, Gérard-Marchant P, Sheppard K, Reddy T, Weckesser W, Abbasi H, Gohlke C, Oliphant TE (2020) Array programming with NumPy. Nature 585(7825):357–362

36. Leontis NB, Westhof E (2001) Geometric nomenclature and classification of RNA base pairs. RNA 7(4):499–512

37. Wiseman T, Williston S, Brandts JF, Lin LN (1989) Rapid measurement of binding constants and heats of binding using a new titration calorimeter. Anal Biochem 179(1): 131–137

Chapter 6

Application of Molecular Dynamics to Expand Docking Program's Exploratory Capabilities and to Evaluate Its Predictions

Wojciech K. Kasprzak and Bruce A. Shapiro

Abstract

Recognition of the growing importance of RNA as a target for therapeutic or diagnostic ligands brings the importance of computational predictions of docking poses to such receptors to the forefront. Most docking programs have been optimized for protein targets, based on a relatively rich pool of known docked protein structures. Unfortunately, despite progress, numbers of known docked RNA complexes are low and the accuracy of the computational predictions trained on those inadequate samples lags behind that achieved for proteins. Compared to proteins, RNA structures generally have fewer docking pockets, have less diverse electrostatic surfaces, and are more flexible, raising the possibility of producing only transiently available good docking targets. We are presenting a docking prediction protocol that adds molecular dynamics simulations before and after the actual docking in order to explore the conformational space of the target RNA and then to reevaluate the stability of the predicted RNA–ligand complex. In this way we are attempting to overcome important limitations of the docking programs: the rigid (fully or mostly) target structure and imperfect nature of the docking scoring functions.

Key words MD simulations, Docking, RNA, Conformational space

1 Introduction

The many already elucidated and emerging cellular functions of RNA have elevated interest in this important molecule. Initially thought of as a transient template for the expression of proteins encoded in the genetic DNA, RNA has been demonstrated to play that role only for approximately 2–3% of the transcripts, while the noncoding majority plays functional roles, including their involvement in disease pathways. As such, RNA is a potential although still challenging and underexplored target for therapeutic agents, such as the short interfering RNAs (*i.e.*, siRNA used against mRNA), engineered RNA nanoparticles (for examples, see [1–5] and references therein), or small molecules [6, 7]. RNA structure is

Jienyu Ding et al. (eds.), *RNA Structure and Dynamics*, Methods in Molecular Biology, vol. 2568,
https://doi.org/10.1007/978-1-0716-2687-0_6,

responsible for the molecule's functionality. Despite advances in the elucidation of RNA structures by experimental methods, the numbers of nucleic acid structures (both DNA and RNA) in the PDB database are just below 3500, approximately half of them being RNA structures, as compared to over 150,000 protein structures [https://en.wikipedia.org/wiki/Protein_Data_Bank, as of April, 2020]. One of the most important sources for RNA structural information on common structural motifs has been the structure of the ribosome. Computational structure prediction and modeling methods benefited enormously from recognition of these motifs that are employed in prediction of new structures when only the primary sequence information is available, as well as in the design of functional RNA-based nanostructures. Characterization of structural dynamics and interactions with ligands that may additionally induce structural and dynamic changes add a level of difficulty to the computational prediction and modeling of RNA structures [8].

The majority of small molecule docking prediction programs have been validated on and optimized for protein targets. The key structural differences between RNA and proteins and the much smaller validation and training sets for known RNA docked complexes result in generally lower accuracy of predictions for RNA targets. RNA structures are generally more flexible and dynamic than proteins, displaying a more even distribution of charge on their surface, and presenting fewer clearly identifiable docking pockets. Among the freely available docking programs, rDock has a separate scoring function, an approximation of the docking free energy, that performs better for RNA targets than for proteins, benefiting mostly from the de-solvation term. The rDock total score is a weighted sum of intermolecular, ligand intramolecular, site intramolecular, and external restraint terms. The main intermolecular component representing the RNA–ligand interaction score includes the solvation potential (while its variant for protein–ligand interaction does not) that in the tests conducted by the authors was shown to be a crucial factor [9]. After extensive testing we have adopted it as our default choice for docking to RNA, but we have developed a protocol that enhances what rDock alone can do, and this protocol is presented below. In short, our approach to the docking prediction problem starts with simulations of the target structure dynamics, on the assumption that the RNA conformations that could yield the best (most stable) docking poses may be transient in nature and lie outside the experimentally captured conformations or computationally predicted models. Next, we perform a large-scale docking screen against a sample of the predicted dynamic RNA conformations. While this protocol is based on the premise that the actual structure of the docked complex is unknown, the whole process can benefit from including whatever experimental data is available to aid in the selection of the predictions that are consistent with it. From high-throughput screening for ligands via a small molecule microarray (SMM) [10, 11], experimentally obtained target (receptor) RNA structures, and NMR and/or X-ray data indicating changes in the individual nucleotide mobility in response to docking, experimental data can be used to validate localization of docking pose predictions, and/or focus the search on the most likely docking site within a larger structure. Finally, the selected complexes are subjected to another round of dynamics simulations to reevaluate their stability. The key protocol steps are illustrated in Fig. 1.

Thus the overall idea is to avoid missing the best docking predictions due to unexplored conformations of the target, and to evaluate best-predicted poses with MD force field parameters that are more accurate than the more approximate binding energy calculations employed by the docking programs.

2 Molecular Dynamics Simulations of RNA Receptors

Because there are many variations in the setup of computing environments and queuing systems, we are assuming that the reader already has a knowledge of his or her local environment. The NIH's Biowulf HPC cluster is our reference system, with the modules package that facilitates initialization of the environment variables necessary to run a given application. If modules are available on your system, look up available packages via a call *module avail* or set up environment variables pointing to the right paths or simply add full paths to the binaries you are invoking.

For example, a call *module load amber/18-gpu* facilitates the running of a GPU-based Amber package on this cluster. A call to initialize *module load amber* is sufficient to run the Amber MD preparation and analysis tools. Other modules used in this protocol are *rDock* and, optionally, *OpenBabel.* We will list Unix shell commands and input parameter files necessary to execute the protocols we are introducing. Submitting them properly as scripts to a queuing system is left to the reader.

The introduction to a sample molecular dynamics protocol presented here works with Amber 16, 18, and 20 packages [12–14], https://ambermd.org/contributors.html and Amber 20 expanded RNA force field choices. We refer the reader to the manual for details. We have previously devoted an entire protocol publication to present details of MD simulations, and so we will only present sample input files with more modern syntax than that used previously and updates relative to Amber 14 employed in it [15]. Rather than explaining all of the numerous parameters employed, we will first summarize the protocol, and then comment on the selected key parameter choices. This step is illustrated in Fig. 1a.

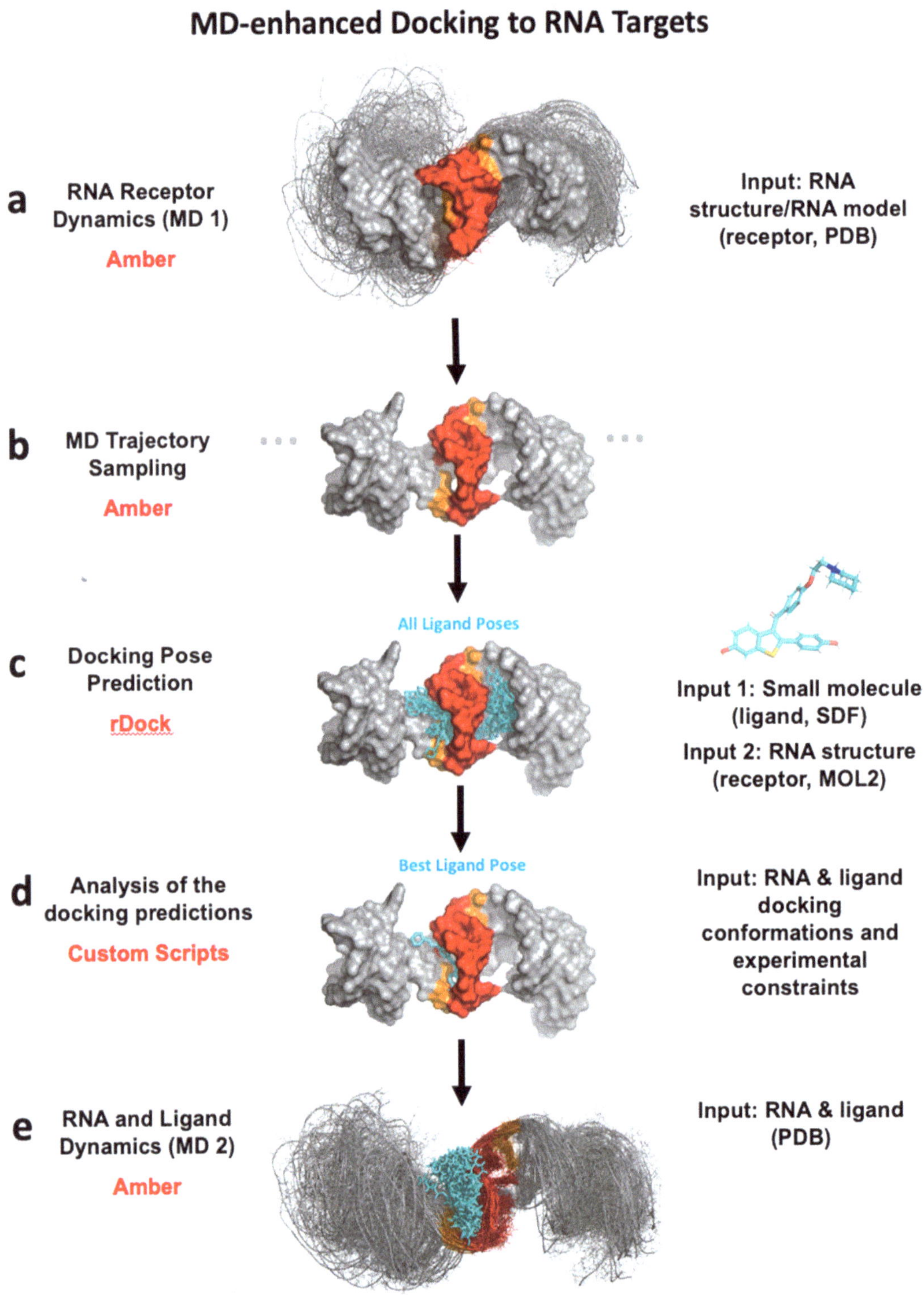

Fig. 1 An illustration of the docking protocol enhanced by the use of molecular dynamics (MD) to explore the RNA target's conformational space as the first step and to evaluate the predicted complex's stability as the last step of the process. In all the 3D illustrations the fragment of the target RNA that proves to be the best docking target (a large bulge loop) is colored red and orange. The illustrated ligand (Raloxifene, a selective estrogen receptor modulator) is colored cyan. (**a**, **b**) Molecular dynamics for an input RNA structure explores

The Amber module called LEaP is used to create the simulated system of RNA (solute) in a solvent box with TIP3P-type waters. The protocol presented here employs RNA force field parameters ff99OL3 (source file leaprc.RNA.OL3), combining parmbsc0 and chi.OL3 parameters for nucleic acid dihedrals [16, 17]. Particle mesh Ewald (PME) dynamics simulations with explicit solvent are employed [18]. A cut-off of 9 Å for the nonbonded interactions was used. The boundary between the solute and solvent box edge is set to at least 12 Å. Explicit solvent periodic boundary conditions are employed. The net RNA charge is neutralized with monovalent Na^+ ions, and additional monovalent salt Na^+/Cl^- ion pairs are added to simulate the net salt concentration of 0.15 M. Simulations are run with 2 fs time steps, except in the equilibration stages where a 1 fs time step is employed. The SHAKE algorithm is used to constrain all hydrogen bonds in the system. The Berendsen thermostat and Berendsen algorithm are used to maintain the simulation temperature of 300 K and the pressure at 1.0 Pa in NPT simulations used in all phases of MD [19].

2.1 Use LEaP to Prepare the System for Simulations

Here are the LEaP commands that can be saved into a LEaP input script file and run via a command *tleap -f tleap-script.in* or executed interactively after invoking variants of the program called tleap or xleap:

```
source leaprc.RNA.OL3
source leaprc.water.tip3p
mol=loadpdb RNA-target.pdb
check mol
savepdb mol RNA-target-leap.pdb
saveAmberParm mol RNA-target-leap.prmtop RNA-target-leap.
inpcrd
addIons mol Na+ 0
solvateBox mol TIP3PBOX 12.0 0.832
addIonsRand mol Na+ NA_NUM Cl- CL_NUM 5
savepdb mol RNA-target-solv.pdb
saveAmberParm mol RNA-target-solv.prmtop RNA-target-solv.
inpcrd
quit
```

Fig. 1 (continued) the conformation space (**a**) and is sampled into a set of individual RNA targets used by the docking program (**b**, only one conformer is shown; gray ellipses indicate multiple samples). (**c**) Docking predictions for a ligand or ligands proceed in parallel for all the generated RNA targets. (**d**) In the docking data analysis phase, the predictions are evaluated based on their docking scores and any experimental information that can be used as additional constraints, and the best docking pose (or several top poses) is selected for the complex stability testing. (**e**) RNA–ligand complex identified in step (**d**) is subjected to MD in order to evaluate its dynamic stability, based on the state-of-the-art MD force fields

(*see* **Note 1**)

The placeholders NA_NUM and CL_NUM in the above LEaP script must be replaced with pre-calculated numbers of ions that yield the target net salt concentration, based on the simplified formula: volume(Å^3) x saltcon(M/l) x 6.022x10^{-4} . The system volume is listed by LEaP following the solvateBox command and the typical concentration used is 0.15 M.

The reason why in the above LEaP script we save the PDB file in addition to Amber parameter files as soon as it passes internal check is to have an Amber-compliant structure file available. RNA-only topology and coordinates files are useful if in the MD simulations one chooses to save the dynamics data only for the solute (RNA), rather than for the entire system (RNA, ions, and water).

2.2 Perform System Equilibration

We employ a multistep equilibration protocol. The first step is energy minimization of the solvent while the RNA is immobilized with strong harmonic restraints. Short phases of heating to 300 K, dynamics at this temperature, and energy minimizations with gradually decreasing harmonic restraints are then applied to the system with the aim of randomizing solvent and gradually exposing the solute to the thermal motions of the solvent around it. Finally, in the last step of the equilibration protocol, with all the restraints on the solute removed, heating to 300 K over 0.2 ns is performed, followed by a constant temperature dynamics run for the balance of 2.0 ns. The production MD simulation is conducted for several hundred nanoseconds to a microsecond or longer, depending on the computational resources and real time limitations. In general, multiple runs with independent equilibrations are recommended as they sample the conformation space of the target RNA better than one very long run. If multiple NMR frames are available as input structures, they should be also considered as starting conformations to independent simulations.

Here are the key input files used in equilibration:

1. System energy minimizations: *minN.in*

```
 MIN solvent, hold 100 nt-long RNA (ntr=1); NVT (ntb=1,ntt=0);
SHAKE (ntc=2);
 Minimization: 1000 steps of steepest decent, then conjugate
gradient for 10000 steps total
 &cntrl
 imin=1,
 ntb=1, ntt=0,
 ntx=1, ntrx=1, ntxo=1, nscm=100,
 irest=0, cut=9.0,
 ntpr=100, ntwx=100, ntwe=100,
 nsnb=10,
```

```
dielc=1.0,
drms=0.01, ntmin=1,
ncyc=1000, maxcyc=10000,
ntr=1, restraint_wt=WEIGHT, restraintmask=':1-100',
&end
```

(*see* **Note 2**)

2. Heating the system: heatN.in

```
HEAT(0.02ns)/MD(till 0.04ns) solvent; hold 100 nt-long RNA
(ntr=1); (Berendsen ntt=1; ntb=1 const. Volume; ntp=1 isotro-
pic; H-bond SHAKE ntc=ntf=2)
Depending on the dt=0.001/0.002 adjust numbers of steps (1ps
is 500@dt=0.002, 1000@dt-0.001)
&cntrl
imin=0, irest=0, ntx=1
ntb=1,
ntp=0, pres0=1.0, taup=0.2, comp=44.6,
ntc=2, ntf=2, tol=0.00001,
cut=9.0
dt=0.001, nstlim=40000,
ntwprt=NUM_ATOMS,
ntpr=1000, ntwx=1000, ntwe=1000,
ntxo=1, ntrx=1,
dielc=1.0, nscm=100, nsnb=10,
ntt=1, tempi=10.0, temp0=300.0,
ibelly=0,
ntr=1, nmropt=1, restraint_wt=WEIGHT, restraintmask=':1-100',
ig=-1,
t=0.0,
&end
&wt
type='TEMP0', istep1=0, istep2=20000, value1=10.0, va-
lue2=300.0,
&end
&wt
type='TEMP0', istep1=20001, istep2=40000, value1=300.0, va-
lue2=300.0,
&end
&wt
type='END',
&end
```

(*see* **Note 3**)

3. Short dynamics simulations: mdpmeN.in

```
MD with holding RNA(ntr=1); Berendsen ntt=1; NPT ntb=2/ntp=1;
H-bond SHAKE ntc=ntf=2.
Write output/ene/traj every ntpr steps (500 at dt=0.002=1ps =
1000@dt=0.001),
for a total of 10,000(dt=0.002)/20,000(dt=0.001) steps = 20ps
= 0.02 ns
&cntrl
imin=0, irest=0, ntx=1,
ntb=2,
ntp=1, pres0=1.0, taup=0.2, comp=44.6,
ntt=1, tempi=100.0, temp0=300.00,
ntc=2, ntf=2, tol=0.00001, tautp=0.2,
cut=9.0,
dt=0.001, nstlim=20000,
ntwprt=NUM_ATOMS,
ntpr=1000, ntwe=1000, ntwx=1000,
ntxo=1, ntrx=1,
dielc=1.0,
nscm=0, nsnb=10,
ibelly=0,
ntr=1, restraint_wt=WEIGHT, restraintmask=':1-100',
ig=-1,
t=0.0,
&end
```

(*see* **Note 4**)

4. The total equilibration sequence of shell commands

Print statements provide comments to the shell commands, and the PRMTOP and INPCRD variables are meant to be set to the topology in coordinate files generated in the initial LEaP session:

```
module load amber/18-gpu
set AMB_MIN = "mpirun -np 1 $AMBERHOME/bin/pmemd.cuda" #
energy minimizations
set AMBER = "mpirun -np 1 $AMBERHOME/bin/pmemd.cuda.MPI" #
heating/dynamics
set PRMTOP = RNA-target-solv.prmtop # your files from LEaP
set INPCRD = RNA-target-solv.inpcrd
printf " Equilibration step 1 of 15: solvent minimization (RNA
restraint = 200.0)\n"
$AMB_MIN -O -i min1.in -o min1.out -p $PRMTOP -c $INPCRD -ref
$INPCRD -r min1.rst
printf " Equilibration step 2 of 15: solvent heating (RNA
```

```
restraints = 200.0)\n"
 $AMBER -O -i heat1.in -c min1.rst -p $PRMTOP -ref min1.rst -o
heat1.out -r heat1.rst -x heat1.x
  printf " Equilibration step 3 of 15: solvent mdpme (RNA
restraints = 200.0)\n"
 $AMBER -O -i mdpme1.in -c heat1.rst -p $PRMTOP -ref heat1.rst
-o mdpme1.out -r mdpme1.rst
  printf " Equilibration step 4 of 15: solvent min2 (RNA
restraint = 25.0)\n"
 $AMB_MIN -O -i min2.in -c mdpme1.rst -p $PRMTOP -ref mdpme1.
rst -o min2.out -r min2.rst
  printf " Equilibration step 5 of 15: solvent mdpme2 (RNA
restraints = 25.0)\n"
 $AMBER -O -i mdpme2.in -c min2.rst -p $PRMTOP -ref min2.rst -o
mdpme2.out -r mdpme2.rst
 ... steps 6 through 12 are minimizations with restraints of
20.0, 15.0, 10.0, 5.0, 2.5, 1.0, 0.1
  printf " Equilibration step 13 of 15: solvent min10 (RNA
restraints = 0.01)\n"
 $AMB_MIN -O -i min10.in -c min9.rst -p $PRMTOP -ref min9.rst
-o min10.out -r min10.rst
  printf " Equilibration step 14 of 15: solvent min11 (NO
restraints)\n"
 $AMB_MIN -O -i min11.in -c min10.rst -p $PRMTOP -ref min10.rst
-o min11.out -r min11.rst
 printf " Equilibration step 15 of 15: solvent and RNA heat2/md
(2.0 ns, no restraints)\n"
 $AMBER -O -i heat2.in -c min11.rst -p $PRMTOP -ref min11.rst
-o heat2.out -r heat2.rst -x heat2.x -inf mdinfo-eql-heat2-
pme.out
```

(*see* **Note 5**)

2.3 Molecular Dynamics Production Run

Following the equilibration, the production run can begin, based either on a one-step full simulation run or a scripted loop with shorter individual stages and restarts. The second option may be more robust, allowing restarts from the last successfully completed stage in case of any failures that may cause the restart file to be corrupted. The production MD run command is shown below, followed by the *md.in* file with parameters set up for a 10 ns-long simulation. The key difference from the equilibration stages is the longer time step of 2 fs (speeding-up execution).

```
echo " Amber 18 MD run with PME protocol"
$AMBER -O -i md.in -p $PRMTOP -c heat2.rst -o md-01.out -r
md-01.rst -x md-01.x -inf mdinfo-01.out
The production MD simulation input: md.in
MD for 5M steps = 10ns; Berendsen ntt=1; ntb=2, ntp=1 isotropic
```

```
(NTP); H-bond SHAKE ntc=ntf=2)
Write all outputs sparsely every 50000 steps = 0.1ns (100 frames
per 10ns; good for long trajectories)
Amber 16/18 binary format netCDF (ioutfm=1, default) for
trajectory (not default in A14) and restart files (ntxo=2)
&cntrl
 imin=0,
 irest=1, ntx=5,
 ntb=2, ntp=1, pres0=1.0, taup=0.5,
 ntt=1, comp=44.6,
 tempi=300.0, temp0=300.0,
 ntc=2, ntf=2, tol=0.00001,
 ntrx=1,
 dt=0.002, nstlim=5000000, ioutfm=1, ntxo=2,
 cut=9.0,
 ntwprt=NUM_ATOMS,
 ntpr=50000, ntwe=50000, ntwr=50000, ntwx=50000,
 nscm=100, nsnb=10, dielc=1.0,
&end
```

(*see* **Notes 6** and **7**)

3 Molecular Dynamics Trajectory Sampling

Following generation of an MD trajectory, we want to sample the conformational space of the RNA receptor by extracting MD states separated in time by 0.5 ns. In our tests we have found that sparser sampling may miss good targets, while denser (more frequent) trajectory sampling yields nearly identical docked poses and docking scores in clusters around the well-scored 0.5 ns sampling states. An example ptraj (cpptraj) input file *ptraj-MD-to-PDBs.in* for trajectory sampling is shown below:

```
# Usage: module load amber; cpptraj -i ptraj-MD-to-PDBs.in
# Option 1: Full solvent trajectory out
parm RNA-target-solv.prmtop [REC]
# Option 2: RNA-only trajectory out
# parm RNA-target-leap.prmtop [REC]
# Wildcards in input file spec are fine, but be careful to
maintain the time sequence
# Trajectory sampling every 10 output frames ex: trajin
md-[0-9][0-9]0.x 0 1000 10 parm [HBV]
# For the above md.in, writing out 100 frames for a 10 ns-long
stage (every 0.1 ns)
# sample down to 0.5 ns snapshots (every 5th sample from the
trajectory)
trajin md-[0-9][0-9]0.x 0 100 5 parm [HBV]
```

```
# For full solvent trajectory, strip solvent, ions, ligand
first, otherwise comment out this section
strip :WAT
strip :Cl-
strip :Na+
# strip :101
# Write out pre-sampled trajectory snapshots PDB frames (sub-
stitute PREFIX)
trajout PREFIX.pdb pdb multi keepext NOBOX
```

(*see* **Notes 8** and **9**)

4 rDock Ligand Docking to RNA Receptors

Once the desired number of RNA target (receptor) conformations is available, the key stage of this protocol, the prediction of docking poses for one or many ligands can be performed on the generated set of receptors (Fig. 1c). Below is a protocol for a standard docking, free of any experimental constraint information with the aid of rDock, in which a flexible ligand is best fit to the defined interface of the rigid pure RNA target (receptor). For more advanced options we refer the reader to the rDock Reference Guide included with the sourceforge distribution file (see https://www.rdock.org, https://sourceforge.net/projects/rdock/) [9].

The docking protocol steps are:

1. Preparation of the ligand file in the Structure Data format.
2. Preparation of the receptor file in the Tripos MOL2 format.
3. Preparation of the docking system definition file, including the receptor and docking cavity parameters.
4. Generation of the docking site parameter file with the aid of the rbcavity rDock's module.
5. Generation of the docking poses using the rbdock module.

4.1 Prepare the Ligand SD File

rDock requires ligands to be in the Structure Data File (SDF) format (.sd, .sdf). These can come from databases, such as *ZINC* (zinc.docking.org). For example, go to the page http://zinc15.docking.org/substances/home/ where you can search for compounds in multiple ways or even draw them. Searching for Raloxifene, for example, will yield the core (ZINC3581478 entry) and the full compound (ZINC538275 entry). Following the full compound link will open the page with detailed chemistry and structure information, and download options as SDF or MOL2. *PubChem* database (https://pubchem.ncbi.nlm.nih.gov) can also be used as a source of ligands. If format conversions are necessary or the 3D structures obtained from a database do not have the explicit hydrogens, the program OpenBabel (http://openbabel.org/wiki/Main_Page) can be employed:

```
module load OpenBabel.
obabel -i mol2 'LIG.mol2' -o sdf 'LIG.sdf' -h
```

The -h adds all hydrogens missing in the input file. The exact protonation state of the ligand has to be determined by the user. Input and output file types can be the same in the above command (*i.e.*, you can augment an SDF input file).

Pure text structural descriptors in the SMILES format can also be translated into 3D structures with the aid of OpenBabel:

```
obabel -i smi 'LIG.smi' -o sdf 'LIG.sdf' --gen3D
```

Please, refer to OpenBabel documentation for all the options associated with conversion to 3D.

Ligands may also be available as X-ray crystal or NMR structures on their own, or as components of complex structures deposited in the PDB database (Protein Data Bank).

One easy way to extract them from the complexes is to open a complex (receptor & ligand) in a visualization program such as PyMOL [https://pymolwiki.org/index.php/Main_Page , https://pymol.org/2/] or UCSF Chimera [https://www.cgl.ucsf.edu/chimera, [20]], select the ligand and save it as an SDF file or PDB or MOL2 file and use OpenBabel to convert it to an SDF format file required by the rDock program.

4.2 Prepare RNA Receptor Files in the Tripos MOL2 Format

In preparation for docking, create a working directory for converting the sampled PDB files to the Tripos (Sybyl) MOL2 file format required by rDock for receptors. While Amber tools LEaP and cpptraj can directly output MOL2 files, the Sybyl format options produce atom types that are different from the posted MOL2 standard and incompatible with rDock.

Amber antechamber module, on the other hand, can be used for that purpose, as well as the freely available program OpenBabel (http://openbabel.org/wiki/Main_Page and https://github.com/openbabel/openbabel/) can also be used.

```
mkdir RECEPTOR-PDB-to-MOL2
cd RECEPTOR-PDB-to-MOL2
Copy (cp) or move (mv) all the sampled PDB files to this
directory:
cp ../PREFIX-*[0-9].pdb .
Option 1: Convert PDB RNA receptor files to the MOL2 format
using antechamber
module load amber
antechamber -i PREFIX.pdb -fi pdb -o PREFIX.mol2 -fo mol2 -c
gas -at sybyl -nc -99 -m 1 -s 2
```

(*see* **Note 10**)

```
Option 2: Convert PDB RNA receptor files to the MOL2 format
using Open Babel
module load OpenBabel.
obabel -i pdb PREFIX-001.pdb -o mol2 -O PREFIX-001.mol2
obabel PREFIX-*.pdb - -o mol2 -m
```

(*see* **Notes 11–13**)

4.3 Prepare the System Definition File with the Receptor and the Docking Cavity Parameters

Create a file with a list of all MOL2 receptor file names to be used by *rDock-PARAMS-SWARM.bash* script in the next step

```
ls -1 PREFIX*.mol2 > mol2list
```

Run the shell script *rDock-PARAMS-SWARM.bash* (listed below) to create rDock parameter files and the NIH Biowulf swarm task scheduling scripts that will most efficiently run in parallel to all the independent docking cavity and ligand docking jobs for each of the RNA target files. Swarm processes a list of commands from a command file and automatically submits them to the batch system for execution. Because in our protocols we are dealing with hundreds or more target conformations of the same molecule and just a few select ligands, in this simple shell script the targets are read from the *mol2list* file, while the reference atom to be used as the center of the docking cavity for all of them and the few ligand names are hardwired (one shown), and must be edited for a specific task before the script execution.

The template file *prm-coords.tmpl* used by the shell script is listed first:

```
RBT_PARAMETER_FILE_V1.00
TITLE gart_DUD
RECEPTOR_FILE MOLNAME
RECEPTOR_FLEX 3.0
#-------------------------------------------------------------
# CAVITY DEFINITION: TWO SPHERE METHOD
#-------------------------------------------------------------
SECTION MAPPER
 SITE_MAPPER RbtSphereSiteMapper
 CENTER (CX,CY,CZ)
 RADIUS 60.0
 SMALL_SPHERE 1.0
 LARGE_SPHERE 6.0
 MIN_VOLUME 100
 MAX_CAVITIES 99
 VOL_INCR 0.0
```

```
 GRIDSTEP 0.8
END_SECTION
#-------------------------------------------------------------
#CAVITY RESTRAINT PENALTY
#-------------------------------------------------------------
SECTION CAVITY
 SCORING_FUNCTION RbtCavityGridSF
 WEIGHT 1.0
END_SECTION
```

(*see* **Note 14**)

```
rDock-PARAMS-SWARM.bash mol2list PREFIX (prm-coords.tmpl has
to be in the same directory)
#!/bin/bash
# Loop through the receptor file list (mol2list), then output
SWARM filed
# for running rbcavity (rDock phase 1: docking cavity genera-
tion) for all MOL2s
# and for rbdock (rDock phase 2: docking ligands to receptors)
files for each ligand
# LIGAND NAMES and CENTER ATOM MUST BE EDITED as indicated in
the script
if [ $# -lt 2 ]; then
 echo "Usage: ./rDock-SWARM.bash <MOL2LIST> <OUTPUT-PREFIX>"
 exit
fi
# Collect command line arguments
pref=$2
echo ""> $pref-cav.swarm
# For all the target file name, generate rbcavity and rbdock
commands
for fname in $(cat ${1}); do
 MOLNAME='basename $fname .mol2'
 printf "For MOL2 file: $fname\n generate parameter file:
$MOLNAME.prm\n"
 cp prm-coords.tmpl $MOLNAME.prm
### CHANGE AS NEEDED the center atom of the search space
(cavity) ##############
# Ex: look for "P.3 17 U" atom definition allowing for variable
numbers of spaces
 XYZcoords='grep 'P.3\s*17\s*U' $fname'
# replace the rbcavity center coords placeholders with center
atom's x,y,z
 printf " Docking cavity central atom: $XYZcoords\n"
 X='echo $XYZcoords | cut -d' ' -f3'
 Y='echo $XYZcoords | cut -d' ' -f4'
 Z='echo $XYZcoords | cut -d' ' -f5'
```

```
# echo $X $Y $Z
 sed -i 's/MOLNAME/'$fname'/g' $MOLNAME.prm
 sed -i 's/CX/'$X'/g' $MOLNAME.prm
 sed -i 's/CY/'$Y'/g' $MOLNAME.prm
 sed -i 's/CZ/'$Z'/g' $MOLNAME.prm
# write out rbcavity command
 printf "rbcavity -r $MOLNAME.prm -was -d\n" >> $pref-cavity.
swarm
### CHANGE THIS SECTION TO MATCH LIGAND(S) TESTED ##########
# -p dock_solv.prm - docking parameters
# -n 50 - run 50 docking attempts for the ligand (-i) to the
receptor cavity parameters (-r)
 printf "rbdock -i LIGAND1.sdf -o $MOLNAME-LIG1 -r $MOLNAME.
prm -p dock_solv.prm -n 50\n" >> $pref-rDock-LIG1.swarm
############################################################
done
cat *.o >> rDock-SWARM.log
```

For example, for a sample input list of three MOL2 input files in a mol2list file (*NMR01.mol2* through *NMR03.mol2*), the center atom and ligands (only one LIGAND1 is shown above) specified in the above script, it will output individual cavity parameter files, based on a template file named prm-coords.tmpl, and swarm files for generating the docking cavities for all the RNA targets and others for the actual docking of all the ligands:

```
NMR01-RNA.prm, NMR02-RNA.prm, NMR03-RNA.prm, PREFIX-cavity.
swarm, and PREFIX-rdock-LIGAND1.swarm
```

4.4 Generate the Docking Site (Cavity) File Using rDock's rbcavity Module

The shell script *rDock-PARAMS-SWARM.bash* executed in step 5.3 has generated the sample script file containing cavity parameter files that can be executed by a queue manager or as individual Unix commands:

```
The PREFIX-cavity.swarm file:
rbcavity -r NMR01-RNA.prm -was -d
rbcavity -r NMR02-RNA.prm -was -d
rbcavity -r NMR03-RNA.prm -was -d
```

In the *rbcavity* command the system definition parameters file (-r) is used to generate the docking site on the RNA target, the -was option instructs the program to output the docking site binary file (extension .as). The -d option (independent of the -was) outputs the cavity volume(s) as a grid file in the InsightII format, which can be used in visualization programs.

4.5 Generate Docking Pose Predictions for the Ligand(s) Using rDock's rbdock Module

The second Swarm input file generated by *rDock-PARAMS-SWARM.bash* contains the docking commands, as shown below for one (*LIGAND1.sdf*) of potentially many ligands:

```
rbdock -i LIGAND1.sdf -o NMR01-RNA-LIG1 -r NMR01-RNA.prm -p
dock_solv.prm -n 50
rbdock -i LIGAND1.sdf -o NMR02-RNA-LIG1 -r NMR02-RNA.prm -p
dock_solv.prm -n 50
rbdock -i LIGAND1.sdf -o NMR03-RNA-LIG1 -r NMR03-RNA.prm -p
dock_solv.prm -n 50
```

As with rbcavity individual commands, rbdock commands shown above can be executed as Linux shell commands in a proof-of-concept test run for just one RNA receptor-ligand.

In the *rbdock* commands, the input (-i) is the ligand file in the MDL SDF format, and the output (-o) specifies the file name for the MDL SD output file. The -r specifies the system definition file (same as in the rbcavity run), and the -p argument specifies the docking protocol parameters file, in this case the scoring function yielding the best results for nucleic acid targets available in $RD_ROOT/data/scripts directory, where RD_ROOT is the rDock installation root path. Finally, the -n argument specifies the number of docking attempts (50) to be made.

(*see* **Notes 15** and **16**)

4.6 rDock Results Analysis

SortAnalyzeSDs.bash script listed below splits SDF output files into single docking results (from the N repeats requested in rbdock calls), sorts the split SDs into subdirectories, writes score reports for each large SD, and records the best scored dock of each RNA target in a "PREFIX-besthits.txt" plain text file.

(*see* **Note 17**)

```
ls -1 *.sd > sdlist
SortAnalyzeSDs.bash sdlist NMRs-rDock
Generated output:
NMRs-rDock-NMR01*0ns-LIG-sorted-scorereport.txt - rank and
detailed score (no orig. indices)
NMRs-rDock-NMR01*0ns-LIG-sorted.sd - matching sorted SD output
NMRs-rDock-besthits.txt - the best results for every input
target
subdirectories: ./NMRs-rDock-allsd - all individual SD sort-
ranked (*-1.sd = best)
 ./NMRs-rDock-bestsd - the best SD files for every input target
rDock post-processing modules:
sdsort sorts SD records by the user requested filed. Output is
written to the standard output and must be redirected to the
output file in the command line.
sdsort -r -f<data_field>
sdfilter processes the output SD file by filtering the records
as specified by the user.
```

```
sdfilter -f <data field><operator><value> [-s <data_field>]
[SDfiles] or
sdfilter -f<ffilename> [-s <data_field>][SDfiles]
```

(*see* **Note 18**)

```
sdreport -t. - produces tab-spaced report for each of the
processed records
The SortAnalyzeSDs.bash script file:
#!/bin/bash
########################################################
# INPUT 1: Text file listing rDock output SD files (with
multiple predicted docking poses),
# one SD file name/line
# INPUT 2: Prefix for besthits file, and sub-directories with
split SDs
# Note: uses rDock utilities sdsort, sdreport and sdsplit
########################################################
if [ $# -lt 2 ]; then
 echo "Usage: ./SortAnalyzeSDs.bash <SDs_list_file> <prefix
for subdir and besthits.txt>"
 exit
fi
# Command line arguments (SDlist=$1)
sddirname=$2
# Create directories for all SD files, and for the best-scored
dock of every run
allsddir=$sddirname-allSDs
bestsddir=$sddirname-bestSDs
mkdir $allsddir
mkdir $bestsddir
# Creates File that will list the scores of the best hits from
each run
echo "RECEPTOR-LIGAND TOTAL INTER INTRA RESTR VDW"\
 > $sddirname-rDock-besthits.txt
# For all the rDock SD output file names, do the processing
(sdraw = fname)
for fname in $(cat ${1}); do
 sdname='basename $fname .sd'
 printf "Working on SD file: $sdname\n"
 # Use sdsort to sort the N docking results in the large SD file
by element "SCORE"
  # -fSCORE sorts on the total score. You can sort by other
useful components such as
  # SCORE.INTER.norm - best for comparing ligands of very
different sizes
 sdsort -n -fSCORE $sdname.sd > $sdname-rDock-sorted.sd
 # Use sdreport to create a report for the all runs per frame
```

```
 sdreport -t $sdname-rDock-sorted.sd > $sdname-rDock-sorted-
scorereport.txt
 # Set bestline to the line containing the highest-scored entry
from *-scorereport
 bestline='head -3 $sdname-rDock-sorted-scorereport.txt | tail
-1 | cut -c35-90'
 # Output scoreinfo for the best line in the besthits file
 printf "$sdname $bestline\n" >> $sddirname-rDock-besthits.txt
 # Use sdsplit to split the sorted SD output file into
individual SDs (-1)
 sdsplit -1 -o$sdname-rDock-rank- $sdname-rDock-sorted.sd
 # Copy best results to "bestsd" directory & move all split SD
files to "allsd"
 cp $sdname-rDock-rank-1.sd ./$bestsddir/
 mv $sdname-rDock-rank-*[0-9].sd ./$allsddir/
done
```

(*see* **Notes 19–21**)

4.7 Molecular Dynamics Simulations of RNA–Ligand Complexes

As we indicated earlier, one can select from all the generated poses the best-scored docking ligand–RNA complexes, optionally considering the poses that best match available experimental constraints (treating docking scores as secondary criteria) for post-docking evaluation in another round of molecular dynamics simulations (Fig. 1d). The MD protocol is nearly identical to the one we outlined in the first step, except that the solute preparation now includes the ligand as well as the RNA receptor (Fig. 1e), and we need to generate a parameter file for the ligand before we can employ LEaP. The Amber antechamber package can be used to generate a MOL2 file for each of evaluated ligands, combining the 3D coordinates from the predicted docking pose with atom types matched to the general Amber force field – GAFF. Antechamber is also used to generate the ligand library PREP file. Then, the Amber utility parmchk2 is employed to generate an FCRMOD file that augments the GAFF data with force field parameters not found there. The GAFF, PREP, FRCMOD, together with the RNA-specific force field parameters are read into LEaP. Finally, the RNA receptor and the ligand PDB files are read-in and combined with TIP3P waters and monovalent ions (Na^+/Cl^-), and topology and coordinate files needed to run the MD simulations are written out. The commands needed to accomplish the steps described above are:

1. Parameter preparation:

Predicted ligand docking poses have to be converted from their SDF format to PDB via OpenBabel (or alternatives, as discussed earlier), and then the (Amber) antechamber package needs to be employed to generate a MOL2 file for each ligand with atom types matched to the Amber general force field – GAFF. Antechamber is also used to produce the ligand library PREP file, while the another Amber utility called parmchk2 produces FCRMOD file containing any force field parameters not listed in GAFF.

```
antechamber -i ligand.pdb -fi pdb -o ligand-ante-bcc.mol2 -fo
mol2 -c bcc -nc 1 -m 1 -s 2
-c bcc: AM1-BCC best method for calculating partial charges
(short of quantum mechanics calculations)
-nc: net charge (specific to the ligand; check in the database
from which it comes or with the chemists who synthesized it for
the protonation state of the ligand).
-m: multiplicity of 1
antechamber -i ligand-ante-bcc.mol2 -fi mol2 -o ligand-ante.
pdb -fo pdb. (optional)
antechamber -i ligand-ante-bcc.mol2 -fi mol2 -o LIGAND-ante-
bcc.prep -fo prepi
parmchk2 -i ligand-ante-bcc.mol2 -f mol2 -o LIGAND-ante-bcc.
frcmod
```

2. *LEaP* commands (execute interactively or from a script file: *tleap -f tleap-script-RNA-LIG.in*)

```
source leaprc.RNA.OL3
source leaprc.gaff
source leaprc.water.tip3p
loadamberprep LIGAND-ante-bcc.prep
loadamberparams LIGAND-ante-bcc.frcmod
REC = loadPdb RNA-target.pdb
check REC
LIG = loadPdb ligand.pdb - could be loadMol2 ligand.pdb
check LIG
COM = combine{REC, LIG}
check COM
addIons COM Na+ 0
check COM
solvateBox COM TIP3PBOX 12.0 0.832
addIonsRand COM Na+ NA_NUM Cl- CL_NUM 5
savepdb COM RNA-LIG-solv.pdb
saveAmberParm COM RNA-LIG-solv.prmtop RNA-LIG-solv.inpcrd
quit
```

(*see* **Notes 22** and **23**)

5 Notes

1. It is generally recommended to examine the input PDB files for consistency of residue and atom naming with the PDB 3.x format. Correctly set up Amber should also be able to recognize and map the older PDB 2.x convention. It is best to let Amber replace and name hydrogens in LEaP, and remove them from the input structure before invoking LEaP. A sample shell command to do so is shown below:

```
egrep -v "[0-9]H|H[A-Z,0-9]|CONECT" $RNA.pdb > $RNA-noH.pdb
```

2. Throughout the equilibration protocol, minimizations are run N times (N=1-11) with decreasing restraint weights (restraint_wt=WEIGHT) set to 200.00, 25.0, 20.0, 15.0, 10.0., 5.0, 2.5, 1.0, 0.1, and 0.01 (kcal/mol.Å^2). The final (11-th) minimization is unrestrained, *i.e.*, the entire last line of parameters starting with ntr=1 should be deleted. The number of minimizations steps can be adjusted depending on the size of the system. Monitor the output file (minN.out) to make sure that the improvements in the total energy of the system level off before the minimization run ends. The sample restraint mask shown (:1-100) applies to all atoms of RNA residues, 1 through 100 (total hypothetical RNA size). It could be substituted with a mask restraining only the heavy atoms (*i.e.*, non-Hydrogens): ':1-100 & !@H='

3. This is a short heating phase from 10 K to 300 K and MD at 300 K to randomize solvent motions. The first heating should be done with the constant Volume setting (ntb = 1, NVT) as the initial adjustments to the voids made in the water box to accommodate the solute and ions may lead to excessive solvent box dimension changes and system crash if we start with the NPT regime. The second heat2.in run will switch to the constant pressure setting (ntp =1, ntb = 2, NTP), which is used throughout the rest of the simulation. Heat1.in is run with the strong WEIGHT of 200.0; the second heat2.in at the end of the full equilibration is restraints-free. The parameter ntwprt shown pointing to a placeholder NUM_ATOMS can be set to 0 (output all atoms of the system) or a number of atoms of the RNA alone, the RNA and ions, or the RNA and ligand when running MD after docking. Unless the entire solvent information is necessary, writing only the dynamics information for selected atoms into output trajectory files is recommended.

   In short, the key differences in *heat2.in* are: ntb=2, ntp=1, nstlim=2000000, no restraints, *i.e.*, no line starting with ntr=1, and the second istep2=2000000 to match the maximum number of simulation steps set in nstlim.

4. WEIGHT equals to 200.0 in mdpme1, then 25.0 in mdpme2. NUM_ATOMS are explained in step 2.
5. System restart files provide starting state information for the next step and are also used as a reference relative to which harmonic restraints are applied.
6. If one adopts iterative MD stages of 10 ns or more in length for which a simulation on the order of 1 μs may take a week for an ~100 nt-long structure, it is also helpful to generate conformation snapshots at the end of each stage for monitoring purposes. This can be accomplished from the restart files as follows:

```
set MDS = "RNA-target-md-01"
cpptraj -p $PRMTOP -y md-01.rst -x $MDS-solvent-box.pdb
cat $MDS-solvent-box.pdb | egrep -v "WAT|Na|Cl" | grep -v -A
1 "TER" | grep -v "\-\-" > $MDS.pdb
```

One has to keep in mind that even if the MD trajectory is written for only the RNA atoms (see NUM_ATOMS in md.in file), the binary restart files contain the entire system information, and that is why we need to strip the solvent and ions from the "raw" PDB file generated from the restart file.

7. The laid-out protocol is not the only valid option for RNA molecular dynamics simulations. It is a safe and tried option, which can be modified by users to include steadily improving Amber RNA force field parameters, water models, divalent ions, thermostat algorithms, or a different selection of the harmonic restraints applied during equilibration. We encourage the reader to check the Amber tutorials available on the Amber group website https://ambermd.org/tutorials/, and try alternative protocols presented in the literature, such as [21, 22].
8. The option multi in the trajout command results in each trajectory frame being written to a separate file with the frame number added to the file name before the extension as follows PREFIX.number.pdb. It may be desirable to change these names with standard padded index numbers of the format PREFIX-###.pdb, in order to have correctly listed numbers 001 through 999 (in this example) with the aid of the Unix command rename:

```
rename PREFIX. PREFIX.0 PREFIX.?.pdb => PREFIX.01.pdb ..
PREFIX.09.pdb
rename PREFIX. PREFIX-0 PREFIX.??.pdb. => PREFIX-001.pdb ..
PREFIX-099.pdb
```

The remaining frames 100 through 999 in this example need not to be changed.

Thus at the end of this process we end up with 2N frames for a trajectory N ns-long, and these are the input files that will be used in the next step as the docking targets.

9. If computational resources are a limiting factor in exploring docking to hundreds of RNA conformations, one can attempt to cluster the sampled trajectory first and use cluster centroids or best energy states in clusters as the first docking targets to be followed by extended searches among the most promising cluster representatives in the RMSD-based or trajectory time-based neighborhoods of those. A lot of factors should be taken into consideration while attempting to cluster the MD solution space in order not to miss distinct conformations, such as the number of clusters requested as a function of the observed levels of collective or localized motions in the simulation (simply watch the trajectory or/and calculate the RMSD relative to its starting conformation to assess the level of distortions in a trajectory). A detailed discussion of the strategies is beyond the scope of this presentation, but a sample K-means clustering command requesting 5 clusters is shown below as a potential replacement or complement to the trajout command shown above:

```
cluster CKm :2-100@P kmeans clusters 5 randompoint \
out PREFIX-2-100P-Km-cnum-v-time.dat \
summary PREFIX-2-100P-Km-clusters-summary.dat \
info PREFIX-2-100P-Km-clusters-info.dat \
repout PREFIX-2-100P-Km repfmt pdb \
avgout PREFIX-2-100P-Km-ave avgfmt pdb
```

10. The key arguments are

-c gast - quick look-up Gasteiger partial charges. There is no need to spend time on more accurate partial charges calculation as these are only placeholders and are not used by rDock.

-at sybyl – Sybyl atom types

-nc -99 - net charge of -99 for a hypothetical 100 nt-long Amber-compatible RNA without the 5′-most phosphate group (H-capped).

11. File format conversions are generally a source of problems. Ultimately, a quick rDock validation run should be performed to make sure all the atom types in the converted MOL2 receptor files are recognized or if a debugging process may be necessary. The call is as follows, and the setup of the input parameter file is explained in the next section:

```
module load rDock
rbcavity -r PREFIX.prm
```

12. In our tests Open Babel versions 2.4.1 and newer yielded rDock-compatible MOL2 files, while the older versions did not. Command "obabel -H" lists all command line options of the program. The first obabel call above converts only one input file. The second variant of the obabel call should be sufficient in most cases. It takes the names of input files and uses them with the .mol2 extensions to store the converted structural information. If further customization of the output MOL2 file names is desired, a simple shell script can be written that iterates over all the input files and writes out MOL2s with custom names.
13. Please, refer to the document, http://help.accelrysonline.com/ulm/onelab/1.0/content/ulm_pdfs/direct/reference/ctfileformats2016.pdf for definitions of the SDF and MOL2 file formats.
14. The most important parameter that has to be selected by the user and entered into the below rDock-PARAMS-SWARM.bash script is the CENTER of the search space, which we base on an atom approximately central to the observed RNA target in the MD conformations. Parameter RADIUS defines the sphere around the CENTER for which the docking cavity search (with the granularity of the GRIDSTEP) will be performed by the rDock module rbcavity. A smaller radius can be employed for a more focused docking around a CENTER that is implicated as a potential docking site by other data. A radius large enough to encompass the entire target structure should be employed in cases of exploratory, "blind" docking. A two-sphere cavity search procedure (see SMALL_SPHERE and LARGE_SPHERE parameters) excludes all the search space (sphere) grid points accessible to the larger of the two probes, while all the search space grid points accessible to the small probe are selected as part of the docking cavity. Once the parameter files PREFIX.prm are generated for the target RNA, the script listed below, a quick run of rbcavity -r PREFIX.prm can be used to see one or multiple cavities identified by the program. Changes to the large and small sphere radii can thus be quickly explored by the user.

15. To use the swarm files as parallel batch job inputs on the NIH Biowulf system, submit the cavity swarm file, followed by rDock swarm files using the SLURM queuing system dependencies and parallelization. For example, submitting a swarm file to generate docking cavities and two swarm files for two hypothetical ligands can be done as follows:

```
swarm -f PREFIX-cavity.swarm --module=rDock
swarm -f PREFIX-rdock-LIGAND1.swarm --module=rDock --depen-
dency afterany:9001
swarm -f PREFIX-rDock-LIGAND2.swarm --module=rDock --depen-
dency afterany:9002
```

Numbers following "afterany;" are hypothetical job IDs listed by the queuing manager after the preceding swarm submissions. The above sequence of calls will run cavity calculations until they finish, followed by the docking calculations for LIGAND1.sdf, then the docking for LIGAND2.sdf. A "ballpark" time for a run of 600 RNA conformations of a 61 nt-long stem-loop target is approximately 1 h, with the runs themselves taking about 40 min each (~20 min for submission for all 600 jobs)

16. Employing the SLURM "dependency" option when submitting jobs helps to avoid exceeding the maximum number of commands allowed in a swarm file or the maximum number of CPUs the user is allotted.

17. This script calls rDock commands *sdsort*, *sdreport*, and *sdsplit*, and it must be run on a platform where the rDock binary is available.

18. Multiple, logically OR'ed filters are allowed and can be specified in filename (in the last call) or one per line. Output is written to standard output and has to be redirected to an output file on the command line.

19. As indicated in the comments to the *sdsort* command in the above script, if one deals with ligands of very different sizes, the intermolecular scores normalized on the number of heavy atoms in the ligands captured in *SCORE.INTER.norm* can be used.

20. For comparative analysis of several ligands, plotting the scoring statistics for all or the best N scores is a good way of finding if the ligands considered are statistically different from the docking prediction perspective. Attention should be paid to the lowest and the highest scores reported in the statistics, as we have observed a small fraction of runs that not only fail to find a pose with a negative score but are stuck in a geometry with van der Waals clashes resulting in out-of-scale positive scores. Such

outliers should be removed from the set of results as they may influence the statistics for a relatively small subset of the top N poses. Another option is to employ *-t SCORE.INTER numeric_threshold -cont* in the rbdock calls, in which case only the docking poses better than the threshold are output to the SD file.

It goes beyond the scope of this protocol outline, but collecting the data on the proximity of the ligand that the nucleotides of the RNA target within some cutoff distance in all the rDock-predicted poses can be plotted as a histogram of sequence index vs frequency. Even a simplified histogram based on the distances from a user-selected center atom of ligand and a heavy atom shared by the nucleotides can be informative. Thresholding this data on the best docking score levels may indicate if there is a topological docking preference as a function of the docking. Optionally, if experimental data on chemical shifts are available this proximity histogram can be used to select a subset of the docking poses that match the experimental pattern.

Since the scoring functions remain to be the least reliable factors in the computational docking predictions, one should keep in mind an option of rescoring the generated poses by independent programs, such as outlined in [23].

21. rDock documentation and tutorials are available from the University of York website at http://www.ysbl.york.ac.uk/rDock/index.html and http://www.ysbl.york.ac.uk/rDock/support.html#tutorials. Support is also offered on the SourceForge site https://sourceforge.net/projects/rdock/support where the latest rDock version is available for download. Finally, further development of rDock has been taken over recently by the RxDock project, which provides video tutorials and additional documentation at https://www.rxdock.org.

22. The RNA target (receptor) and ligand files must come from the correct docking pose (*i.e.*, be compatible in 3D). Alternatively, *COM = loadpdb RNA-ligand-complex.pdb* can be used, if the complex is available in one file, instead of the separate REC and LIG *loadpdb* and the *combine* commands shown above.

    Following creation of the system via the above *LEaP* command, one can use the same MD protocol described for the RNA simulations, except that the harmonic restraints that were previously applied to residues 1-100 of the hypothetical RNA receptor, should now be extended to include the ligand that is seen as the 101st "residue." All other notes apply.

23. In our recent study we have employed an optional step that goes beyond the scope of this protocol and that involved performing quantum mechanics calculations to obtain more accurate partial charges than those that can be generated by

antechamber and splicing them into the ligand MOL2 file. Packages such as GAMESS [24] or Gaussian [25] can be used for that purpose. Any experimental information about divalent ions placement should be included, as Mg^{2+}, for example, may be structurally important, by coordinating certain tertiary interactions. Such coordinating ions should be included in the restraints list in the equilibration steps.

Acknowledgment

This research was supported by the Intramural Research Program of the NIH, National Cancer Institute, Center for Cancer Research (to B.A. Shapiro) and with Federal funds from the Frederick National Laboratory for Cancer Research, National Institutes of Health, HHSN261200800001E (to W.K. Kasprzak). We would also like to acknowledge Brigit Shields and Nour Ali Ahmed for help with developing the protocol described here. Computational support by the NCI Advanced Biomedical Computational Science group (ABCS), as well as the NIH Helix/Biowulf facility, is highly appreciated. The content of this publication does not necessarily reflect the views or policies of the Department of Health and Human Services, nor does mention of trade names, commercial products, or organizations imply endorsement by the U.S. Government.

References

1. Afonin KA, Kasprzak WK, Bindewald E et al (2014a) In silico design and enzymatic synthesis of functional RNA nanoparticles. Acc Chem Res 47:1731–1741
2. Afonin KA, Viard M, Koyfman AY et al (2014b) Multifunctional RNA nanoparticles. Nano Lett 14:5662–5671
3. Afonin KA, Viard M, Kagiampakis I et al (2015) Triggering of RNA interference with RNA-RNA, RNA-DNA, and DNA-RNA nanoparticles. ACS Nano 9:251–259
4. Parlea L, Puri A, Kasprzak W et al (2016) Cellular delivery of RNA nanoparticles. ACS Comb Sci 18:527–547
5. Zakrevsky P, Kasprzak WK, Heinz WF et al (2020) Truncated tetrahedral RNA nanostructures exhibit enhanced features for delivery of RNAi substrates. Nanoscale 12:2555–2568
6. Meyer SM, Williams CC, Akahori Y et al (2020) Small molecule recognition of disease-relevant RNA structures. Chem Soc Rev 49: 7167–7199
7. Disney MD (2019) Targeting RNA with small molecules to capture opportunities at the intersection of chemistry, biology, and medicine. J Am Chem Soc 141:6776–6790
8. Kasprzak WK, Ahmed NA, Shapiro BA (2019) Modeling ligand docking to RNA in the design of RNA-based nanostructures. Curr Opin Biotechnol 63:16–25
9. Ruiz-Carmona S, Alvarez-Garcia D, Foloppe N et al (2014) rDock: a fast, versatile and open source program for docking ligands to proteins and nucleic acids. PLoS Comput Biol 10: e1003571
10. Connelly CM, Abulwerdi FA, Schneekloth JS Jr (2017) Discovery of RNA binding small molecules using small molecule microarrays. Methods Mol Biol 1518:157–175
11. Labuda LP, Pushechnikov A, Disney MD (2009) Small molecule microarrays of RNA-focused peptoids help identify inhibitors of a pathogenic group I intron. ACS Chem Biol 4:299–307

12. Salomon-Ferrer R, Case DA, Walker RC (2013) An overview of the Amber biomolecular simulation package. Wiley Interdiscip Rev Comput Mol Sci 3:198–210
13. Case DA, Cheatham TE, Darden T et al (2005) The Amber biomolecular simulation programs. J Comput Chem 26:1668–1688
14. Case DA, Ben-Shalom IY, Brozell SR et al (2018) Amber 2018. University of California, San Francisco, CA
15. Kim T, Kasprzak WK, Shapiro BA (2017) Protocols for molecular dynamics simulations of RNA nanostructures. Methods Mol Biol 1632:33–64
16. Perez A, Marchan I, Svozil D et al (2007) Refinenement of the AMBER force field for nucleic acids: improving the description of alpha/gamma conformers. Biophys J 92: 3817–3829
17. Zgarbova M, Otyepka M, Sponer J et al (2011) Refinement of the Cornell et al. Nucleic acids force field based on reference quantum chemical calculations of glycosidic torsion profiles. J Chem Theory Comput 7:2886–2902
18. Darden T, York D, Pedersen L (1993) Particle Mesh Ewald – an N.Log(N) method for Ewald Sums in large systems. J Chem Phys 98:10089–10092
19. Berendsen HJC, Postma JPM, Vangunsteren WF et al (1984) Molecular-dynamics with coupling to an external bath. J Chem Phys 81: 3684–3690
20. Pettersen EF, Goddard TD, Huang CC et al (2004) UCSF Chimera – a visualization system for exploratory research and analysis. J Comput Chem 25:1605–1612
21. Kuhrova P, Mlynsky V, Zgarbova M et al (2019) Improving the performance of the amber RNA force field by tuning the hydrogen-bonding interactions. J Chem Theory Comput 15:3288–3305
22. Mlynsky V, Kuhrova P, Kuhr T et al (2020) Fine-tuning of the AMBER RNA force field with a new term adjusting interactions of terminal nucleotides. J Chem Theory Comput 16: 3936–3946
23. Stefaniak F, Bujnicki JM (2021) AnnapuRNA: a scoring function for predicting RNA-small molecule binding poses. PLoS Comput Biol 17:e1008309
24. Schmidt MW, Baldridge KK, Boatz JA et al (1993) General atomic and molecular electronic structure system. J Comput Chem 14: 1347–1363
25. Frisch MJ, Trucks, GW, Schlegel, HB et al (2016) Gaussian 16 Rev. C.01. Wallingford

Chapter 7

Exploring the Binding Process of Cognate Ligand to *Add* Adenine Riboswitch Aptamer by Using Explicit Solvent Molecular Dynamics (MD) Simulation

Lei Bao and Yi Xiao

Abstract

Riboswitches are RNA-structured elements that modulate gene expression by changing their conformation in response to specific metabolite ligand binding. Therefore, the biological functions of riboswitches mainly depend on the switching of secondary and three-dimensional structures in the presence and absence of the metabolite ligands. However, the binding mechanisms of cognate ligands to riboswitches are still not well understood. Here, we have introduced how to use explicit solvent molecular dynamics (MD) simulation to observe the binding process of cognate ligand to *add* adenine riboswitch aptamer at the atomic level. In addition, we have analyzed the driving factors of the binding process and calculated the binding free energy based on the Molecular Mechanics Poisson-Boltzmann Surface Area (MM-PBSA) method.

Key words *Add* adenine riboswitch, Molecular dynamics, Ligand recognition, Binding free energy

1 Introduction

Riboswitches are ligand-responsive regulatory elements located in the untranslated regions of messenger RNAs. In a suitable ionic environment, they change their conformations in response to specific metabolite binding and achieve the regulatory function of gene expression [1–3]. In the process of investigating the regulatory function of riboswitches, the key point is to understand the ligand recognition mechanism. Although a large number of experiments have explored the ligand recognition process of riboswitches in ionic solution from the perspective of phenomenology [4–9], a deep understanding of the microscopic mechanism of this process at the atomic level is still lacking.

Recently, with the rapid development of in silico techniques, it has become a reality to study the dynamic properties of riboswitches at the atomic level using explicit solvent molecular dynamics (MD) simulations [10–15]. Here, we present a detailed protocol

Jienyu Ding et al. (eds.), *RNA Structure and Dynamics*, Methods in Molecular Biology, vol. 2568,
https://doi.org/10.1007/978-1-0716-2687-0_7,

for modeling the binding process of cognate ligand to *add* adenine riboswitch aptamer in an ionic solution environment using molecular dynamics package Assisted Model Building with Energy Refinement (AMBER) [16] and other relevant tools. Our goal is to help others quickly apply the skill of manipulating MD software to simulate biomolecular systems.

2 Materials

The most important resource for MD simulation is a high-performance computer, especially a Graphics Processing Unit (GPU) server. In addition, we must also have a series of professional simulation software and operating environment. Take the simulation of *add* adenine riboswitch as an example; the materials required will be listed in the following sections.

2.1 Hardware Preparation

In order to understand molecular functions of biomolecules, it is important to sufficiently sample their conformational space. However, long-time simulation is very time-consuming, and using a high-performance computer for MD simulation can greatly shorten the sampling period. Therefore, choosing the appropriate CPU and GPU is the basis of a successful simulation process. In this example, nodes with Intel Xeon E5–2640@2.5GHz CPUs and NVIDIA GeForce GTX 1080Ti GPUs are selected for our explicit solvent MD simulation.

2.2 Software Configuration

1. Operating system: Ubuntu is a modern, open source operating system based on Linux for the enterprise server, desktop, cloud, and so on, which is widely used in the field of scientific research. In this example, Ubuntu 18.04 long time support version is chosen as our operating system.
2. Job scheduling system: Simple Linux Utility for Resource Management (Slurm) is an open source, fault-tolerant, and highly scalable cluster management, and it is used for job management of our MD simulations.
3. Quantum chemical calculation software: Gaussian 09 package [17] is used for the calculations of charge distribution of small organic molecules, and GaussView 16 is used for generating the input files for Gaussian calculations.
4. Molecular dynamics package: AMBER 16 can be downloaded from the website of the AMBER Home Page (http://ambermd.org/). AMBER 16 has been specially optimized for using NVIDIA GPUs to accelerate explicit solvent PME simulations, so it is very suitable for exploring the dynamics of biomolecules in solution. In practice, in order to use CPU

multithreading parallel computing and GPU acceleration technology, AMBER 16 was compiled with MPICH 3.2.1 and CUDA 8.0 in GCC 6.3.0 environment on our hardware.

5. Molecular graphics software: Visual Molecular Dynamics (VMD) 1.9.4 [18] can be downloaded from the website of Theoretical and Computational Biophysics Group at the University of Illinois Urbana-Champaign (https://www.ks.uiuc.edu/), and it is used for displaying, animating, and analyzing large biomolecular systems using three-dimensional graphics and built-in scripting.
6. Other tools: Gedit 3.28.1 and Gnuplot 5.2 are used for text editing and two-dimensional data plotting, respectively. Adobe Photoshop CS 5.0 is used for combining multiple pictures and outputting high-quality images.

3 Methods

3.1 Construction of Initial System

1. Download the initial coordinate file (containing the information of both aptamer and ligand) of *add* adenine riboswitch (PDB ID: 1Y26) [19] from RCSB Protein Databank website (http://www1.rcsb.org/). Then, check the PDB file and delete the unused information in it, leaving only the atomic coordinates. Next, the PDB file is divided into two new PDB files: Apt.pdb for aptamer information and Lig.pdb for ligand information.
2. The parameters of all kinds of residues in nucleic acids are preset by AMBER force field. However, many simulation processes involve the ligand molecules. These small organic molecules vary, and their force parameters and electrostatic information may not be found in the AMBER library file. It is necessary to calculate and generate templates according to specific needs. Fortunately, the built-in antechamber module of AMBER can generates small molecule templates with general AMBER force field (GAFF) [20]. In addition, the charge distribution can be accurately determined by using the antechamber module combined with third-party quantum chemical calculation software, *i.e.*, Gaussian package [17].

   In this example, to generate the force field parameters for the ligand (adenine molecule), we should follow the steps below:

   Step 1. Import the coordinate file Lig.pdb into GaussView and generate the Gaussian input file Lig.gjf according to the basis set of HF/6-31G*.

   Step 2. To optimize the structure with Gaussian, type:

```
g09 < Lig.gjf > Lig.out
```

Step 3. After Gaussian calculation is completed, based on the output file Lig.out, we can fit the Restrained Electrostatic Potential (RESP) charge [21] for the ligand molecule using the antechamber module in AMBER with the following command:

```
antechamber -i Lig.out -fi gout -o Lig_resp.mol2 -fo mol2 -pf
y -c resp
```

Then, we can get the file Lig_resp.mol2 containing the RESP charges of all atoms of the ligand.

Step 4. Last, we need to convert the file Lig_resp.mol2 into the topology parameter file for the ligand. The calculation processes are as follows:

```
antechamber -i Lig_resp.mol2 -fi mol2 -o Lig_resp.prep -fo
prepi
parmchk -i Lig_resp.prep -f prepi -o Lig_resp.frcmod
```

3. Open Gedit text editor and copy the coordinate information of ligand molecule from Lig.pdb into the tail of Apt.pdb. Then, open VMD software and load the Apt.pdb, then, we can view and edit the units of ligand and aptamer with the built-in tools of VMD software. Here, our aim is to study the binding process of free ligand molecules to the aptamer of *add* adenine riboswitch. Thus, in the initial state, we need to keep a certain distance (set to at least 5 Å in this example) between ligand molecules and aptamer, which is easily achieved with the molecule-move tool of VMD. So far, we have completed the loading of the first free ligand molecule. In order to reproduce a realistic environment and accelerate the binding process, it is necessary to place more ligand molecules around the aptamer. Therefore, we repeat the above process several times, and put a total of ten free ligand molecules around the aptamer in this example. At last, a complex system (saved as complex_Apt_Lig.pdb) is generated as the initial structure using VMD software.
4. Load all force field parameters with AMBER and construct the initial system for subsequent MD simulations. To accomplish this task, we need to select the appropriate force field parameter for each element in the molecular system, and load them one by one with the built-in teLeap module in AMBER according to the following steps:

Step 1. To load the parameter for *add* adenine riboswitch aptamer using the RNA force field *ff99OL3* [22, 23] (*see* **Note 1**), open teLeap module and type:

```
source leaprc.RNA.OL3
```

Step 2. Load the parameters for solvent and ions, using the commands below:

```
source leaprc.water.tip3p
loadamberparams frcmod.ionsjc_tip3p
loadamberparams frcmod.ions234lm_hfe_tip3p
```

Here, the solvent model used is the TIP3P water model [24], which is a popular choice in biomolecular simulations (*see* **Note 2**). The parameters for monovalent ions (such as K^+ and Cl^-) come from the work by Joung and Cheatham in 2008 [25], and Li/Merz ion parameters for +2 to +4 ions in TIP3P water [26] is chosen to reproduce the dynamics for high-valence metal ions (such as Mg^{2+}).

Step 3. To load the parameters for ligand molecules, the files Lig_resp.frcmod and Lig_resp.mol2 generated previously are used by issuing two commands:

```
loadamberparams Lig_resp.frcmod
LIG=loadmol2 Lig_resp.mol2
```

Step 4. Load the complex system containing both aptamer and ten ligand molecules, and create a truncated octahedron of water around it:

```
model=loadpdb complex_Apt_Lig.pdb
solvateoct model TIP3PBOX 12.0
```

A truncated octahedron is a shape which has higher spherical symmetry than a cube, thus wasting less computing power on solvent molecules distant from the solute (*see* **Note 3**). In this example, the minimum distance between any atom originally present in solute and the edge of the periodic box is 12 Å, and the resulting solvation box will be repeated in all three spatial directions.

Step 5. From the output of the previous step, we can get the information about the volume of the solvation box. Then, the number of ions to be added can be estimated according to this information. Here, we hope that the system will contain about 0.1 M $MgCl_2$ (*see* **Note 4**) on the basis of neutralization by K^+ ions. As a result, 70 K^+, 35 Mg^{2+}, and 70 Cl^- should be added to the molecular system (*see* **Note 5**). The detailed operations are given as follows:

```
addIonsRand model K+ 70
addIonsRand model MG 35
addIonsRand model Cl- 70
```

So far, the construction of the initial molecular system is basically completed. By checking the output information of teLeap module, we find that the whole system contains 50,426 atoms.

Step 6. Create the force field parameter file complex_solv.prmtop and coordinate file complex_solv.inpcrd for the entire solvated system, and quit the teLeap interface:

```
saveamberparm model complex_solv.prmtop complex_solv.inpcrd
quit
```

By loading these two files into VMD software, we can view and check the initial molecular system (Fig. 1). If there are no problems, we may proceed to the next section, which introduces running minimization and MD simulation.

3.2 Running Minimization and Molecular Dynamics Simulation

3.2.1 Minimization of the Initial System

In the previous section, we built the initial system (containing the coordinate file complex_solv.inpcrd and force field parameter file complex_solv.prmtop) with teLeap module and VMD software. However, due to possible overlaps and conflicts of the atomic coordinates, these two files cannot be used directly for MD simulation at present. To make the structure of the initial system more reasonable physically (corresponding to the actual minima in the force field), it is necessary to minimize the system before running the regular MD simulation.

In this example, the energy minimization of the system will be carried out using Simulated Annealing with NMR-Derived Energy Restraints (SANDER) module in two stages. First, we will hold the *add* adenine riboswitch aptamer and ligand molecules fixed, and just minimize the ions and water molecules. The settings of our first input control file min1.in (*see* **Note 6**) are given as follows:

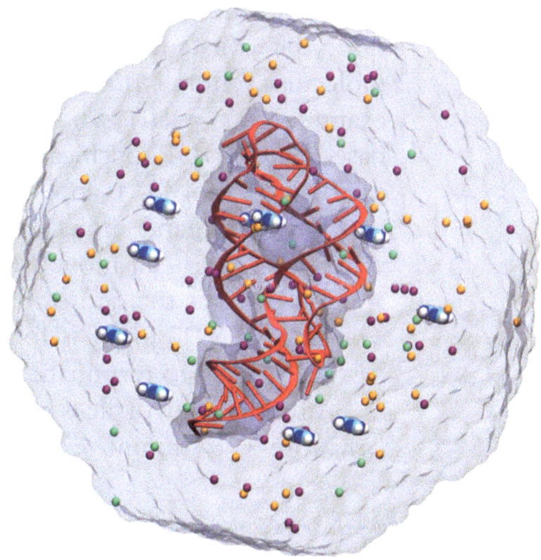

Fig. 1 Snapshot of initial molecular system displayed in the interface of VMD software. The *add* adenine riboswitch aptamer is rendered by the newcartoon coloring method (in red). Ten ligand molecules around the aptamer are shown in the van der Waals coloring method, and the K^+, Mg^{2+}, Cl^- ions are represented by orange, green, and purple balls, respectively

Minimization of ions and solvent:

```
&cntrl
 imin = 1,
 maxcyc = 6000,
 ncyc = 3000,
 ntb = 1,
 ntr = 1,
 cut = 10,
&end
```

Hold the aptamer and ligand molecules fixed:

```
500.0
RES 1 81
END
END
```

The contents between flag *&cntrl* and *&end* specify the detailed setting used for calculation with the SANDER module. To run minimization, *imin* must be set to 1. Since we just want to move atoms away from the local maxima, 6000 time steps of minimization (*maxcyc* = 6000) is enough to achieve our goal. Here, to make the system reach equilibrium more efficiently, we will run the same number of time steps with the steepest algorithm and conjugate gradient method by setting *ncyc* = 3000. In addition, the constant volume periodic boundaries is used for the molecular system (*ntb* = 1), and a cutoff of 10 Å is chosen in this example (*cut* = 10). *ntr* = 1 means that the position restraints based on GROUP specification will be used. Here, a force constant of 500 kcal/mol/Å^2 is applied to restrain residues 1 to 81 (aptamer and ten free ligand molecules), which means that only ions and water molecules can move freely. It is worth noting that "*END*" appears twice at the end of file min1.in, but this is not a mistake. The first "*END*" input is for the current group, and the second "*END*" is used to terminate all groups input.

After the control file is edited, we can use the SANDER module (*see* **Note** 7) to minimize the system for the first time:

```
mpirun -np 16 sander.MPI -O -i min1.in -o min1.out -c complex_-
solv.inpcrd -p complex_solv.prmtop -r min1.rst -ref complex_-
solv.inpcrd
```

We give a summary of the command line options for the SANDER module as follows:

```
-O Overwrite output files if they exist
-i min1.in Read the input control file
-o min1.out Generate output file of user readable state info
-c complex_solv.inpcrd Read the input initial coordinates and
periodic box size
-p complex_solv.prmtop Read the input molecular topology,
force field and so on
-r min1.rst Generate output final coordinates, velocities for
restart
-ref complex_solv.inpcrd Read the reference coordinates for
position restraints
```

The option "-ref" at the end of the command line is not required normally. It is used for the restrained simulations to map the residue selections specified in the input control file. In general, for simplicity, we can directly use the initial coordinates file as the reference coordinates file.

In this task, 16 CPU threads were called by MPI to speed up the minimization of the system, and it still took about 8 min, so this is a good time to take a short break.

The end of calculation means that the minimization of ions and water molecules has been completed. Then, we will proceed to minimize the entire system as a whole. Similar to the last minimization process, we first create a control file named min2.in:

Minimization of the entire system as a whole:

```
&cntrl
 imin = 1,
 maxcyc = 6000,
 ncyc = 3000,
 ntb = 1,
 ntr = 0,
 cut = 10,
&end
```

The content of min2.in is almost the same as that of min1.in, except that the constraints on aptamer and ligand molecules are removed. Based on this control file, restart file min1.rst and force field parameter file complex_solv.prmtop, we can perform the second stage minimization by typing:

```
mpirun -np 16 sander.MPI -O -i min2.in -o min2.out -c min1.rst
-p complex_solv.prmtop -r min2.rst
```

This minimization will also take several minutes. When the calculation is completed, we can use the following command to load the system into the VMD software for viewing and checking:

```
vmd -parm7 complex_solv.prmtop -netcdf min2.rst
```

3.2.2 Equilibration of the Molecular System

From the previous stage of minimization, we obtain a coordinate file min2.rst in which the geometric positions of all atoms are relatively reasonable. Afterwards, a regular simulation for this molecular system can be started. However, for each MD simulation, all atoms of the molecular system need to undergo a relaxation that usually lasts for at least several nanoseconds (ns) before the system reaches a stationary state. The initial nonstationary segment of the simulated trajectory is often called the equilibration stage.

The equilibration of the molecular system usually can be divided into two stages: (1) heating up the system and (2) equilibration at constant temperature (300 K) and pressure (1 bar). Thus, following this protocol, we will first heat up our molecular system from 0 K to 300 K over 100 picoseconds (ps) of simulation time.

In order to ensure that there are not any wild fluctuations in our solutes (containing aptamer and ligand molecules), we will use a weak constraint on them, as we did in the first stage of minimization. The process of heating up will be carried out with the volume held constant, and the control file heat.in is given as follows:

Heating up the system:

```
&cntrl
 imin = 0,
 ntx = 1, irest = 0,
 ntpr = 1000, ntwx = 1000, ntwr = 1000, iwrap = 1, ioutfm = 1,
 ntt = 3, temp0 = 300.0, tempi = 0.0, ig = -1, gamma_ln = 1.0,
 ntb = 1, ntc = 2, ntf = 2, cut = 10.0,
 nstlim = 100000, dt = 0.001,
 ntr = 1, restraint_wt = 10.0, restraintmask = ":1-81",
&end
```

The meaning of each term is explained line by line below:

`imin = 0: Run molecular dynamics without any minimization.`

ntx = *1*, *irest* = *0*: We are starting a new MD simulation instead of restarting an existing one. Therefore, velocities in the input coordinate file will be ignored, and the time step count will be set to 0.

ntpr = *1000*, *ntwx* = *1000*, *ntwr* = *1000*, *iwrap* = *1*, *ioutfm* = *1*: We will write the energy information (*ntpr*), print the coordinates (*ntwx*) in NetCDF format (*ioutfm*) to the output file every 1000 time steps (1 ps), and save a restart file (*ntwr*), ensuring that recovery from a crash will not be so painful, every 1000 time steps. In addition, the coordinates written to the output and restart files will be placed to a primary simulation box (*iwrap* = *1*), which makes the resulting structures look better visually.

ntt = *3*, *temp0* = *300.0*, *tempi* = *0.0*, *ig* = *−1*, *gamma_ln* = *1.0*: This line is mainly used for the temperature regulation. The molecular system is gradually heated up from the initial temperature 0 K (*tempi*) to the reference temperature 300 K (*temp0*) under the control of Langevin dynamics with a collision frequency of 1.0 ps^{-1} (*gamma_ln* = *1.0*). Then the system should be maintained at 300 K by adjusting the kinetic energy. Here, the random seed is set to be based on the current date and time; thus, it will be different for each simulation.

ntb = *1*, *ntc* = *2*, *ntf* = *2*, *cut* = *10.0*: The constant volume periodic boundaries is used. The SHAKE algorithm [27] should be turned on to constrain bonds involving hydrogen and omit those interactions (*see* **Note 8**). Meanwhile, a cutoff is set to 10 Å to determine the maximum distance for nonbonded interactions (*see* **Note 9**).

nstlim = *100,000*, *dt* = *0.001*: We will run 100,000 time steps with a time step of 1 femtosecond (fs) for the heating process, which gives a total simulation time of 100 ps.

ntr = 1, restraint_wt = 10.0, restraintmask = ":1–81": The aptamer and ligand molecules (residues 1 to 81) will be restrained using the harmonic potential with a force constant of 10 kcal/mol/Å^2, and the reference coordinates are read from the restart file min2.rst generated in the previous minimization stage.

After all the input files are prepared, we can start the simulation of heating up the molecular system. Since MD simulation is a time-consuming process, we will use Slurm to schedule GPU devices to accelerate our calculation from now on, which requires a job script file.

Here, the job script file heat.sh is constructed by using the text editor Gedit. We can open Gedit and type in the following:

```
#!/bin/bash
#SBATCH -J Heating_up
#SBATCH -p gpu
#SBATCH -N 1
#SBATCH -n 1
#SBATCH --gres=gpu:1
#SBATCH --cpus-per-task=1
#SBATCH --ntasks-per-node=1
pmemd.cuda -O -i heat.in -o heat.out -c min2.rst -p complex_-
solv.prmtop -r heat.rst -x heat.ncdf -ref min2.rst
```

In this script, we use Slurm to allocate one GPU and one CPU core in the same node for Particle Mesh Ewald Molecular Dynamics (PMEMD) module performing the simulation. At present, PMEMD has become the primary engine for running molecular dynamics simulations with AMBER. The command line options for PMEMD are exactly the same as the SANDER module. Compared with the SANDER module, however, it can greatly reduce the time required to run implicit and explicit solvent MD simulations because of the ability to utilize NVIDIA GPUs.

The simulation task can be started by submitting the job script file heat.sh with the following command:

```
sbatch heat.sh
```

Thanks to the powerful capability of NVIDIA GeForce GTX 1080Ti GPU, the heating process takes only about 3 min to complete. Next, we will equilibrate the system using pressure and temperature control to adjust the density of water to experimental values. To avoid the abnormal deformation of solutes (especially for the aptamer), we will gradually remove the constraint on them in the equilibration process of the molecular system.

In detail, the equilibration process is divided into three stages with different constraints on solutes: (1) consistent with the heating process (10 kcal/mol/Å^2); (2) reduce the constraints (1 kcal/

mol/Å^2); (3) remove all constraints on solutes. Thus, the contents of the three control files (equil_10.in, equil_1.in and equil_0.in) are identical except for the restraint items. Take the file equil_10.in as an example:

Equilibration of the system (with constraint of 10 kcal/mol/Å^2):

```
&cntrl
 imin = 0,
 ntx = 5, irest = 1,
 ntpr = 10000, ntwx = 10000, ntwr = 10000, iwrap = 1, ioutfm =
1,
 ntt = 3, temp0 = 300.0, tempi = 300.0, ig = -1, gamma_ln = 1.0,
 ntb = 2, ntc = 2, ntf = 2, cut = 10.0,
 ntp = 1, pres0 = 1.0, taup = 2.0,
 nstlim = 1000000, dt = 0.002,
 ntr = 1, restraint_wt = 10.0, restraintmask = ":1-81",
&end
```

Here, since the equilibration process will continue from the previous simulation (heating up of the system), we will use the velocity information from the restart file by setting the items *ntx* and *irest* to 5 and 1, respectively. Additionally, the equilibration process will be carried out with the constant pressure periodic boundary (*ntb* = *2*), and an average pressure of 1 bar (*pres0* = *1.0*) will be maintained by using isotropic position scaling with a relaxation time of 2 ps (*taup* = *2.0*).

Each equilibration stage will be run 1,000,000 time steps with a time step of 2 fs (gives a simulation time of 2 ns). Thus, the total simulation time of the equilibration process is 6 ns. We will start the three MD simulations with the following script file equil.sh:

```
#!/bin/bash
#SBATCH -J Equilibration
#SBATCH -p gpu
#SBATCH -N 1
#SBATCH -n 1
#SBATCH --gres=gpu:1
#SBATCH --cpus-per-task=1
#SBATCH --ntasks-per-node=1
pmemd.cuda -O -i equil_10.in -o equil_10.out -c heat.rst -p
complex_solv.prmtop -r equil_10.rst -x equil_10.ncdf -ref
heat.rst
pmemd.cuda -O -i equil_1.in -o equil_1.out -c equil_10.rst -p
complex_solv.prmtop -r equil_1.rst -x equil_1.ncdf -ref
equil_10.rst
pmemd.cuda -O -i equil_0.in -o equil_0.out -c equil_1.rst -p
complex_solv.prmtop -r equil_0.rst -x equil_0.ncdf
```

After about one and a half hours, the simulation of equilibration is over. We can use VMD software to watch what happens during the equilibration process. If everything looks okay, we are going to perform the production MD simulation for the whole system.

3.2.3 Production MD Simulation of the System

At this stage, we will continue to simulate the system with the same protocol as in the final stage of equilibration until we are satisfied. The control file md.in is as follows:

Production MD simulation of the system:

```
&cntrl
 imin = 0,
 ntx = 5, irest = 1,
 ntpr = 10000, ntwx = 10000, ntwr = 10000, iwrap = 1, ioutfm =
1,
 ntt = 3, temp0 = 300.0, tempi = 300.0, ig = -1, gamma_ln = 1.0,
 ntb = 2, ntc = 2, ntf = 2, cut = 10.0,
 ntp = 1, pres0 = 1.0, taup = 2.0,
 nstlim = 600000000, dt = 0.002,
&end
```

Similar to the previous equilibration stage, we use a job script file md.sh to start the production MD simulation:

```
#!/bin/bash
#SBATCH -J Production_md
#SBATCH -p gpu
#SBATCH -N 1
#SBATCH -n 1
#SBATCH --gres=gpu:1
#SBATCH --cpus-per-task=1
#SBATCH --ntasks-per-node=1
pmemd.cuda -O -i md.in -o md.out -c equil_0.rst -p complex_-
solv.prmtop -r md.rst -x md.ncdf
```

Since the binding of ligand to aptamer may be a long process, we set up a large number of time steps (*nstlim* = *600,000,000*) to ensure that the system is fully sampled (*see* **Note 10**). This will be very time-consuming in computation. Even if GPU is used to accelerate the simulation, the whole calculation still takes about 2 weeks.

3.3 Trajectory Analysis and Binding Free Energy

3.3.1 Trajectory Analysis

Now, we have a complete trajectory of 1.2 microseconds to explore what happens to our molecular system during the dynamic simulation process. In fact, there are many phenomena and parameters that can be analyzed for dynamic processes of biomolecular systems. As a protocol, we only choose the most important phenomenon in our simulated system to investigate, which is the binding process of cognate ligand to *add* adenine riboswitch aptamer.

Usually, after a long MD simulation, some atoms of the biomolecule may migrate outside of the original box. Therefore, the simulation trajectory should be fixed by reimaging using the Cpptraj module for better visualization. The processing script file re_image.in is given below:

```
parm complex_solv.prmtop
trajin md.ncdf
autoimage
rms first :1-71
trajout md_imaged.ncdf netcdf
```

In this script file, the statement "*autoimage*" can help us center and image the trajectory with periodic boundaries automatically, and the statement "*rms first :1-71*" aligns the aptamer in our trajectory with a reference structure (the first conformation) to remove the overall translational and rotational motion.

We use this script file through the following command:

```
cpptraj -i re_image.in
```

After the trajectory is processed, we can visualize the trajectory now with the VMD program by typing:

```
vmd -parm7 complex.prmtop -netcdf md_imaged.ncdf
```

Fortunately, we have observed that one free ligand molecule is successfully captured by the binding pocket of the aptamer in our trajectory. In order to show the process clearly, three typical conformational snapshots are generated using VMD software and extracted in high quality using Photoshop (Fig. 2a).

Moreover, for each simulation system, the properties we want to study may be totally different. Thus, trajectory analysis is a highly custom job. Here, as an example of using the Cpptraj module for analysis, we have calculated the fluctuation of Solvent-Accessible Surface Area (SASA) with LCPO algorithm [28] for the ligand molecule that finally binds to the aptamer. Through VMD visualization, we determine that the residue number of this ligand molecule is 80. Hence, the content of Cpptraj script file sasa.in can be written as below:

```
parm complex_solv.prmtop
trajin md_imaged.ncdf
surf :80 out sasa.dat
```

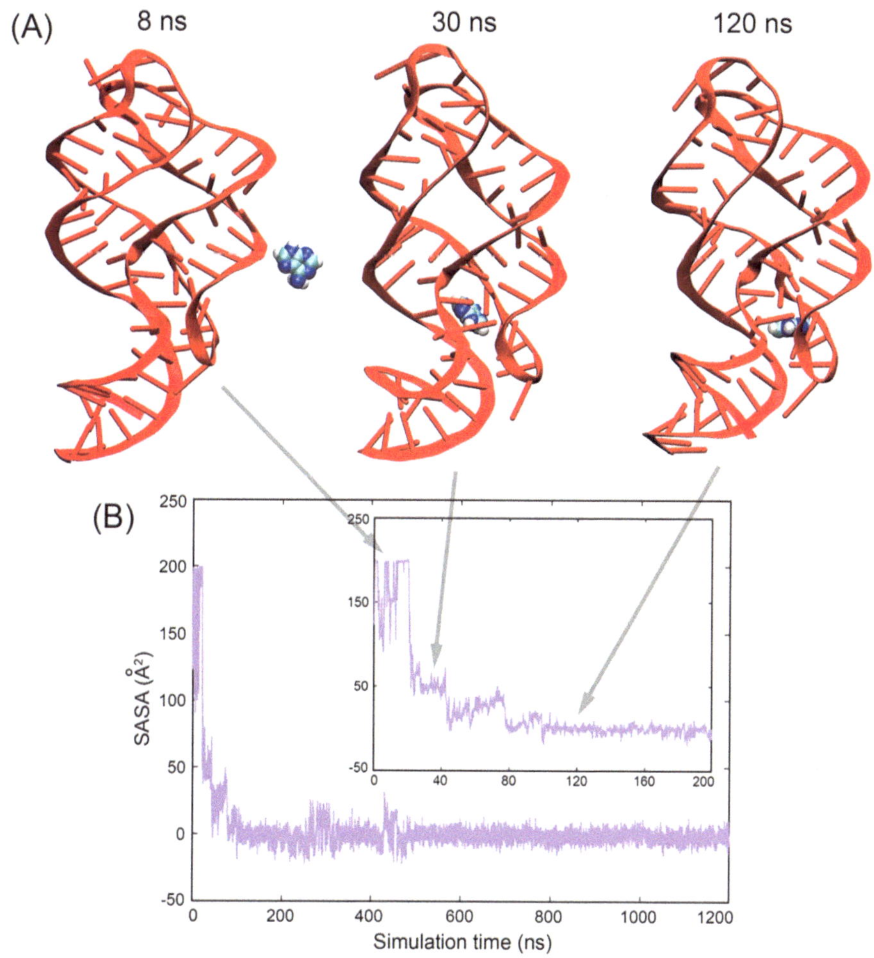

Fig. 2 (**a**) Three typical conformational snapshots for the binding process of a free ligand molecule to *add* adenine riboswitch aptamer at time points of 8 ns, 30 ns, and 120 ns, respectively. (**b**) The fluctuation of SASA for the ligand molecule (finally captured by the binding pocket) versus MD simulation time

We submit this script file using the command "*cpptraj -i sasa.in*" and an output file sasa.dat is obtained. By visualizing the result with Gnuplot (Fig. 2b), we find that the SASA of a diffusive ligand molecule is about 200 $Å^2$. Lying in the groove formed by junctions J2–3 and J3–1 can dramatically reduce the SASA of the ligand molecule to about 50 $Å^2$. If the ligand molecule is lucky enough to get into the binding pocket, it will be completely dehydrated (SASA ≈ 0).

MD trajectories contain abundant useful information. Only when we are diligent in thinking, can we dig out what we want.

3.3.2 Calculation of Binding Free Energy

In this section, we will demonstrate how to calculate the binding free energy of the ligand molecule to *add* adenine riboswitch aptamer with AMBER based on the Molecular Mechanics Poisson-Boltzmann Surface Area (MM-PBSA) method [29].

Firstly, we use a Python utility ante-MMPBSA.py to create compatible complex, receptor, and ligand topology files from the initial topology file complex_solv.prmtop:

```
ante-MMPBSA.py -p complex_solv.prmtop -s ":WAT,K+,Cl-,MG,LIG&!
(:80)" -c complex.prmtop -n ":LIG" -l ligand.prmtop -r recep-
tor.prmtop
```

Secondly, we should generate a trajectory file corresponding to the topology file complex.prmtop with the script file gen_traj.in:

```
parm complex_solv.prmtop
trajin md.ncdf
autoimage
strip :WAT,K+,Cl-,MG,LIG&!(:80)
trajout complex.ncdf netcdf
```

Thirdly, an input file mmpbsa.in needs to be constructed to set all parameters for both PB calculation and entropy estimation with normal mode analysis:

Input file for PB calculation and nmode analysis:

```
&general
 startframe = 30000, endframe = 60000, interval = 1000,
&end
&pb
 istrng = 0.15,
&end
&nmode
 nmode_istrng = 0.15,
&end
```

Here, we have selected 30 evenly spaced snapshots along the last 600 ns production trajectory (binding free energy almost reaches equilibrium) for PB and entropy calculations, and these two calculations are carried out under a typical ionic strength (150 mM) of the physiological environment. Finally, we perform the calculation interactively and print the progress to a log file progress.log by typing following command:

```
mpirun -np 16 MMPBSA.py.MPI -O -i mmpbsa.in -cp complex.prmtop
-rp receptor.prmtop -lp ligand.prmtop -y complex.ncdf > pro-
gress.log
```

Our calculation results show a favorable enthalpy ΔH of about -24.0 kcal/mol and an unfavorable entropic contribution ($-T\Delta S \approx 18.2$ kcal/mol) at the temperature of 300 K. Hence, the total

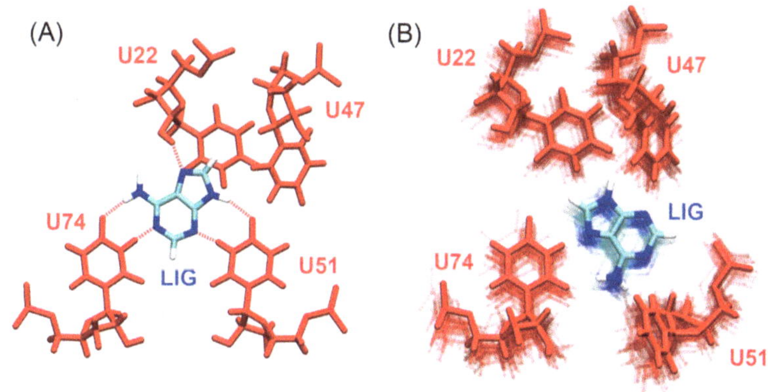

Fig. 3 (**a**) Top view of the binding pocket in the crystal structure of *add* adenine riboswitch (PDB ID: 1Y26). Hydrogen bonds are indicated by red dashed lines. (**b**) Structural ensembles of the ligand molecule in the binding pocket through our simulation, which are generated by taking 30 evenly spaced snapshots along the trajectories (after 600 ns)

binding free energy of our system is about −5.8 kcal/mol, which is in quantitative agreement with previous experimental measurements ($\Delta G \approx -8.8$ kcal/mol) [30]. At the same time, we find that the configuration of the ligand molecule in the binding pocket has not yet reached the minimum energy state (Fig. 3), which may be the reason for the discrepancy of binding free energy between our results and the experiment.

4 Notes

1. This force field has been specially optimized and improves the distribution of an important backbone angle χ, which prevents the duplex structure of RNA from turning into a ladder-like configuration [23].
2. As a long tradition, TIP3P is the default choice for solvent models. In particular, many older force fields were parameterized in simulations using TIP3P as the solvent, and errors in the solvent part of the total energy are compensated to an extent by fitted parameters of the gas phase part. Therefore, replacing TIP3P with other water models without re-optimizing the underlying gas phase force field may not improve the accuracy of the biomolecular simulation.
3. Compared with cuboid box, using a truncated octahedron can save a lot of computing resources. In this example, if the aptamer is hydrated in a cuboid box (keeping a minimum distance of 12 Å from each face of the solvation box), 26,426 water molecules will be added to fill this box. However, only 15,948 water molecules are needed to hydrate the aptamer in a

truncated octahedron. Thus, for our molecular system, choosing the truncated octahedron can increase the computational efficiency by nearly 40%.

4. Due to the high electronegativity of RNA molecules, the distribution of metal ions in the solvation box is uneven. The concentration of metal ions in the region around RNA is significantly higher than that at the edge of the solvation box. It should be noted that even if we ensure that the overall concentration of Mg^{2+} ions in the solvation box is 0.1 M, the bulk concentration of Mg^{2+} ions must be lower than this value. Therefore, it is necessary to pay attention to this issue when comparing the results of MD simulation with the corresponding experiments.
5. Since Mg^{2+} ions are randomly added, some of them may be very close to RNA structure. However, as high-valence metal ions, these Mg^{2+} ions can form stable direct interactions with RNA structure before being fully hydrated, which may cause the system to be trapped in a local minimum and affect the subsequent sampling. Therefore, we need to adjust the positions of Mg^{2+} ions that are too close to RNA structure (using previous method for adding ligand molecules), so that they can be fully hydrated in the equilibration stage of MD simulation.
6. For a complicated MD simulation, it is very important to have a good naming system for the files we create. Having a good naming system can make our work easy to look back and use elsewhere. In contrast, a bad naming system can result in a lot of unnecessary troubles in the future.
7. With the continuous update of AMBER software, minimization is currently supported for single GPU runs with the PMEMD module. In this example, we still use the SANDER module to minimize the molecular system in a CPU node. Additionally, it should be noted that the parallelized executable "*sander.MPI*" may be inefficient on some systems if the number of CPU processors is not a power of 2.
8. Since explicit solvent simulations are significantly more computationally expensive than implicit solvent simulations, it is critical to reducing computational complexity as much as possible. One way is to use a special "three-point" water model [31] (such as TIP3P) and remove the bond stretching freedom in it. Consequently, a larger time step (2 fs) can be used in MD simulations and the bond interactions involving H-atoms are no longer considered, which can greatly improve the computational efficiency of explicit solvent simulations.
9. Due to certain aspects of the GPU pair list operation, it is unsafe to run MD simulations that are less than three times

the nonbonded cutoff in any given direction of the solvation box, and these cases will be trapped with an error message until a fix can be implemented.

10. Due to the limitation of computational power, MD simulation cannot fully sample the molecular system in a limited time through a single trajectory. As a result, the binding process of cognate ligand to *add* adenine riboswitch aptamer we observed may be just an accidental event. To validate the statistical significance of this phenomenon, the simplest way is to do multiple simulations under the same conditions. If this event occurs repeatedly in a similar way, the intrinsic mechanism is worthwhile to explore further. In addition, it is a good choice to solve this problem by using enhanced sampling algorithms, such as Replica Exchange Molecular Dynamics (REMD) [32], Gaussian Accelerated Molecular Dynamics (GaMD) [33] and so on.

Acknowledgments

This work was supported by the National Natural Science Foundation of China Grant (No. 11874162) and the Starting Research Fund from Hubei University of Medicine (No. 2020QDJZR015).

References

1. Roth A, Breaker RR (2009) The structural and functional diversity of metabolite-binding riboswitches. Annu Rev Biochem 78:305–334
2. Mandal M, Breaker RR (2004) Gene regulation by riboswitches. Nat Rev Mol Cell Bio 5: 451–463
3. Serganov A, Nudler E (2013) A decade of riboswitches. Cell 152:17–24
4. Suddala KC, Wang J, Hou Q, Walter NG (2015) Mg^{2+} shifts ligand-mediated folding of a riboswitch from induced-fit to conformational selection. J Am Chem Soc 137: 14075–14083
5. Lin JC, Thirumalai D (2008) Relative stability of helices determines the folding landscape of adenine riboswitch aptamers. J Am Chem Soc 130:14080–14081
6. Neupane K, Yu H, Foster DA, Wang F, Woodside MT (2011) Single-molecule force spectroscopy of the *add* adenine riboswitch relates folding to regulatory mechanism. Nucleic Acids Res 39:7677–7687
7. Lemay JF, Penedo JC, Tremblay R, Lilley DM, Lafontaine DA (2006) Folding of the adenine riboswitch. Chem Biol 13:857–868
8. Stagno JR, Liu Y, Bhandari YR, Conrad CE, Panja S, Swain M et al (2017) Structures of riboswitch RNA reaction states by mix-and-inject XFEL serial crystallography. Nature 541:242–246
9. Dalgarno PA, Bordello J, Morris R, St-Pierre P, Dube A, Samuel ID et al (2013) Single-molecule chemical denaturation of riboswitches. Nucleic Acids Res 41:4253–4265
10. Priyakumar UD, MacKerell AD Jr (2010) Role of the adenine ligand on the stabilization of the secondary and tertiary interactions in the adenine riboswitch. J Mol Biol 396:1422–1438
11. Sharma M, Bulusu G, Mitra A (2009) MD simulations of ligand-bound and ligand-free aptamer: Molecular level insights into the binding and switching mechanism of the *add* A-riboswitch. RNA 15:1673–1692
12. Hayes RL, Noel JK, Mohanty U, Whitford PC, Hennelly SP, Onuchic JN et al (2012) Magnesium fluctuations modulate RNA dynamics in the SAM-I riboswitch. J Am Chem Soc 134: 12043–12053
13. Villa A, Wöhnert J, Stock G (2009) Molecular dynamics simulation study of the binding of

purine bases to the aptamer domain of the guanine sensing riboswitch. Nucleic Acids Res 37:4774–4786

14. Bao L, Wang J, Xiao Y (2019) Dynamics of metal ions around an RNA molecule. Phys Rev E 99:012420. 1–19
15. Bao L, Wang J, Xiao Y (2019) Molecular dynamics simulation of binding process of ligand to add adenine riboswitch aptamer. Phys Rev E 100:022412. 1–11
16. Case DA, Cheatham TE, Darden T, Gohlke H, Luo R, Merz KM et al (2005) The Amber biomolecular simulation programs. J Comput Chem 26:1668–1688
17. Frisch MJ et al (2009) Gaussian 09, Revision D.01. 2009
18. Humphrey W, Dalke A, Schulten K (1996) VMD: visual molecular dynamics. J Mol Graph 14:33–38
19. Serganov A, Yuan YR, Pikovskaya O, Polonskaia A, Malinina L, Phan AT et al (2004) Structural basis for discriminative regulation of gene expression by adenine- and guanine-sensing mRNAs. Chem Biol 11: 1729–1741
20. Wang J, Wolf RM, Caldwell JW, Kollman PA, Case DA (2004) Development and testing of a general amber force field. J Comput Chem 25: 1157–1174
21. Bayly CI, Cieplak P, Cornell W, Kollman PA (1993) A well-behaved electrostatic potential based method using charge restraints for deriving atomic charges: the RESP model. J Phys Chem 97:10269–10280
22. Pérez A, Marchán I, Svozil D, Šponer J, Cheatham TE, Laughton CA et al (2007) Refinement of the AMBER force field for nucleic acids: Improving the description of α/γ conformers. Biophys J 92:3817–3829
23. Zgarbová M, Otyepka M, Šponer J, Mládek A, Banáš P, Cheatham TE et al (2011) Refinement of the cornell et al. nucleic acids force field based on reference quantum chemical calculations of glycosidic torsion profiles. J Chem Theory Comput 7:2886–2902
24. Jorgensen WL, Chandrasekhar J, Madura JD, Impey RW, Klein ML (1983) Comparison of simple potential functions for simulating liquid water. J Chem Phys 79:926–935
25. Joung IS, Cheatham TE (2008) Determination of alkali and halide monovalent ion parameters for use in explicitly solvated biomolecular simulations. J Phys Chem B 112:9020–9041
26. Li P, Roberts BP, Chakravorty DK, Merz KM (2013) Rational design of particle mesh Ewald compatible Lennard-Jones parameters for +2 metal cations in explicit solvent. J Chem Theory Comput 9:2733–2748
27. Ryckaert JP, Ciccotti G, Berendsen HJ (1977) Numerical integration of the Cartesian equations of motion of a system with constraints: molecular dynamics of n-alkanes. J Comput Phys 23:327–341
28. Weiser J, Shenkin PS, Still WC (1999) Approximate atomic surfaces from linear combinations of pairwise overlaps (LCPO). J Comput Chem 20:217–230
29. Miller BR, McGee TD, Swails JM, Homeyer N, Gohlke H, Roitberg AE (2012) MMPBSA.py: an efficient program for end-state free energy calculations. J Chem Theory Comput 8: 3314–3321
30. Gilbert SD, Stoddard CD, Wise SJ, Batey RT (2006) Thermodynamic and kinetic characterization of ligand binding to the purine riboswitch aptamer domain. J Mol Biol 359: 754–768
31. Miyamoto S, Kollman PA (1992) SETTLE: an analytical version of the SHAKE and RATTLE algorithm for rigid water models. J Comput Chem 13:952–962
32. Mitsutake A, Sugita Y, Okamoto Y (2001) Generalized-ensemble algorithms for molecular simulations of biopolymers. Biopolymers 60:96–123
33. Miao Y, Feher VA, McCammon JA (2015) Gaussian accelerated molecular dynamics: unconstrained enhanced sampling and free energy calculation. J Chem Theory Comput 11:3584–3595

Chapter 8

Prediction of Dynamic RBP–RNA Interactions Using PrismNet

Wenze Huang and Qiangfeng Cliff Zhang

Abstract

A capacity to detect the binding profiles of RNA targets for an RNA-binding protein (RBP) under different cellular conditions is essential to understand the functions of the RBP in posttranscriptional regulation. However, the prediction of RBP binding sites in vivo remains challenging. Tools that predict RBP–RNA interactions using sequence and/or predicted structures cannot reflect the exact state of RNA in vivo. PrismNet, which uses both sequences and in vivo RNA structure information from probing experiments, can accurately predict RBP binding under different cellular conditions by deep learning, and can be applied for functional studies of RBPs. Here, we provide a detailed protocol showing how to train a PrismNet model of RBP–RNA interactions for an RBP, and how to apply the model for predictions of the RBP binding under different conditions.

Key words RNA-binding protein, RNA structure, RBP–RNA interaction, Deep learning

1 Introduction

RNA-binding proteins (RBPs) play essential regulatory roles in posttranslational regulation of gene expression, including in RNA translation and metabolism [1–4]. To deepen the understanding of RBP-mediated regulatory mechanisms, the first step is determining the profiles of RBP binding sites on RNAs. Technologies have been developed to characterize sequence preferences of an RBP in vitro [5–7] or to profile RBP binding sites in vivo [8–11]. Computational methods have then been developed that use the knowledge garnered from experimental studies for prediction of RBP–RNA interactions [12–15].

RBP–RNA interactions are very dynamic and thus frequently change across different cellular conditions [1, 16, 17]. However, it is expected that prediction tools based on RNA sequence and/or predicted structure (which is essentially also from RNA sequences) cannot reflect the changes of RBP binding in discrete cell

Jienyu Ding et al. (eds.), *RNA Structure and Dynamics*, Methods in Molecular Biology, vol. 2568,
https://doi.org/10.1007/978-1-0716-2687-0_8,

conditions. Studies have found that in vivo RNA structure is also dynamic, which is dependent on and also affects RBP binding [16, 18, 19]. This observation suggests that we may use in vivo RNA structure for prediction of dynamic RBP–RNA interactions in cells.

We thus designed PrismNet, a deep learning network that integrates the sequence and in vivo structure information of binding sites, to model and predict RBP–RNA interactions [16]. PrismNet was trained as a binary classification model for each RBP, with RNA sequences, in vivo RNA structures (icSHAPE scores), and the profile of the RBP binding sites as the input for the model (the in vivo RNA structures and RBP binding profiles are in the same cell conditions). PrismNet uses multiple convolutional layers and residual blocks to learn the sequence and structural pattern covering a distance in RNA molecules. As a convolutional layer contains multiple channels to learn different aspects of RBP-binding, PrismNet uses a "squeeze-and-excitation" module to learn channel-wise attention between convolutional channels. PrismNet further applies a SmoothGrad technique to generate saliency maps, which calculates the contribution of each nucleotide for the RBP binding on a binding site. In this way, PrismNet can help identify the exact locations of binding nucleotides in each (usually broad) RBP binding site. Studies showed that PrismNet can accurately predict RBP binding in vivo and can distinguish the different bindings in discrete cell conditions [16].

Here, we explicate the protocol for the training of a PrismNet model, and the use of it for RBP–RNA binding predictions. We also show how we can analyze a trained model to reveal the RNA sequence and structure integrative motif for the binding of an RBP on RNA.

2 PrismNet Install

PrismNet can be accessed at https://github.com/kuixu/PrismNet. To get started, users first should install Python (Version 3.6), PyTorch (version 1.1.0, with NVIDIA CUDA Support) and the pip tool. Users can clone PrismNet from the GitHub website:

```
$ git clone https://github.com/kuixu/PrismNet.git
```

PrismNet trains a model to capture RBP binding patterns on RNA targets, which could be used for prediction of a new binding site. But before installing a PrismNet model, users should install some prerequisite Python packages (SciPy, termcolor, h5py, pandas, scikit-learn, tensorboardX, tqdm, matplotlib, einops) by running the following commands:

```
$ cd PrismNet
$ pip install -r requirements.txt
```

Then to install all existing PrismNet models in our library, users can run

```
$ pip install -e .
```

3 Data Preparation and Generation

To train a new PrismNet model for one RBP, users should first generate the training dataset which contains experimental CLIP-seq data for the RBP. Briefly, if available, the 5000 most confident binding sites from the CLIP-data will be used as the positive samples (true binding sites). Another 10,000 regions will be randomly sampled from the whole transcriptome, avoiding positive binding regions, as the negative samples (false binding sites). Finally, we randomly split the positive and negative samples into the training dataset and the validation dataset by a certain ratio, *e.g.*, 4:1.

We have pre-compiled the training data for 168 RBPs, which is available at our website (http://prismnet.zhanglab.net/):

```
$ cd PrismNet/data
$ wget https://zhanglabnet.oss-cn-beijing.aliyuncs.com/prism-
net/data/clip_data.tgz
$ tar zxvf clip_data.tgz
```

The input data (*.tsv* format file, Table 1) for each RBP PrismNet model contains six columns. The first column indicates the sample type of data, "A" for training data and "B" for testing data. The second column is the transcript coordinate of the binding sites, split by "|". Users can add other information for samples in this column. The third and fourth columns contain the sequence and icSHAPE score (split by ","). The fifth column contains the binding scores, which are normalized scores of the peaks signal intensities in the CLIP data file. The binding scores in the negative samples are set to −1. Note that the binding scores are not used as the input of model, which is the sixth column, the label column that shows the sample label for the data, with 1 for a positive sample (a true binding site) and 0 for a negative sample (a false binding site).

Again, we have prepared the training and validation datasets for the 168 RBPs, which could be converted to the h5 format files (such as *TIA1_Hela.h5*) by the following command. PrismNet uses

Table 1
Example data of the input file

Sample type	Transcript coordinate	Sequence	icSHAPE score	Binding score	Label
B	ENST00000389680|76|176	AC...A	0.585,0.113,...,0.63	0.98	1
A	ENST00000367142|244|344	GC...C	0.124,0.252,...,1.0	0.72	1
B	ENST00000372686|877|977	TT...A	0.592,0.476,...,0.41	−1	0
A	ENST00000400561|319|419	TT...G	0.372,0.124,...,0.0	−1	0

this file format as the input file format for the training, evaluation, or inference tasks.

```
$ cd PrismNet
$ bash tools/gdata_bin.sh
```

4 PrismNet Model Training and Prediction

4.1 Training

PrismNet uses the binary cross entropy loss as the loss function and the optimizer Adam to optimize the model parameters. The area under the receiver operating characteristic curve (AUC) and the confusion matrix are calculated to evaluate the trained model performance. Technically, to train a PrismNet model of RBP–RNA interactions for an RBP, users can run:

```
$ bash exp/prismnet/train.sh TIA1_Hela clip_data
```

where *TIA1_Hela* is the name of the data file, which can be replaced with the name of the user's data file of interest. *clip_data* is the data file directory. Hyper-parameters can be tuned in *exp/prismnet/train.sh*. For available training options, please look at *tools/main.py*.

The output files of this command include the trained model parameter for TIA1_Hela:

```
models/TIA1_Hela_ics/out/models/TIA1_Hela_PrismNet_pu_best.
pth
```

the AUC and a confusion matrix of the TIA1_Hela model:

```
models/TIA1_Hela_ics/out/evals/TIA1_Hela_PrismNet_pu.metrics
```

and the predicted binding probability for the validation sample.

```
models/TIA1_Hela_ics/out/evals/TIA1_Hela_PrismNet_pu.probs
```

4.2 Evaluation

For evaluation of the trained PrismNet models, users can run the following command:

```
$ bash exp/prismnet/eval.sh TIA1_Hela clip_data
```

where *TIA1_Hela* is the name of the trained model, which should be replaced with the name of the data of interest, and *clip_data* is the data file directory.

4.3 Inference

PrismNet predicts the binding probability of new samples based on a trained model, which can be run using:

```
$ exp/prismnet/infer.sh TIA1_Hela /path/to/inference_file.tsv
```

where *TIA1_Hela* is the name of the trained model, which should be replaced with the name of the data of interest, and */path/to/inference_file.tsv* is the file containing the samples needed to be predicted (the file format is the same as *.tsv file in Table 1). This command will output a file in the text format (*TIA1_Hela.txt*) that contains the binding probability of the samples.

5 Saliency Map Generation

To visualize the binding preference of an RBP to each nucleotide, PrismNet calculates a saliency map to capture both the sequence and structural response signals. The nucleotides in each binding site that are highly responsive, or important for RBP binding, are highlighted in the saliency map, with the quantification of the contribution from every position to RBP binding (Fig. 1).

5.1 Saliency Map Calculation

For calculating the saliency values of saliency maps, we provide the script *saliency.sh*. Users can run it as:

```
$ exp/prismnet/saliency.sh TIA1_Hela /path/to/inference_file.
tsv
```

Here *TIA1_Hela* is the name of the trained model. */path/to/inference_file.tsv* is the file containing the samples to compute the saliency maps.

This command will output a saliency map file *TIA1_Hela.sal* (Table 2). For every five values of "response value" column in the file, the four response values correspond to the four nucleotides (A, C, G, U) and the last response value corresponds to structure (icSHAPE score).

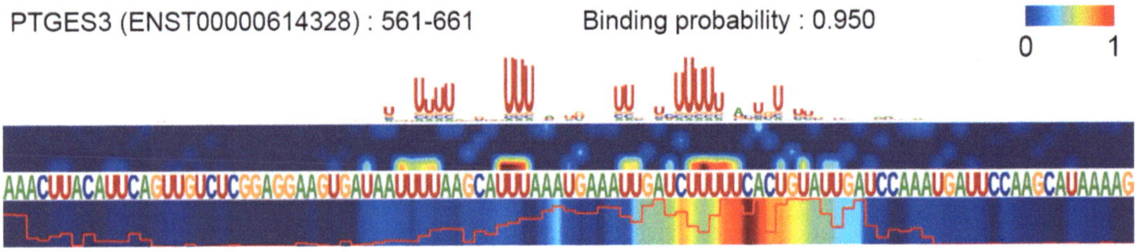

Fig. 1 Saliency map of binding site calculated by TIA1 PrismNet model. The middle is the sequence information for the binding site. The heatmap tracks show the response signal derived from the model at each nucleotide position, while the red color indicates the high attention region (upper heatmap show sequence response and bottom heatmap shows structure response). The logos on the top display the sequence component in the high attention region. The red line on the bottom heatmap represents icSHAPE scores

Table 2
The data format for the saliency file

Sample id	Label	Binding probability	Sequence	icSHAPE score	Response value
0	1	1	U...C	1.0,0.542,...,0	0.0,0.0,...,0.151
1	1	1	A...U	0.037,0.0,...,0.121	0.02,0.0,...,0.01
2	1	0.952	A...G	0.613,0.941,...,0.063	0.01,0.0,...,0.026
3	0	0.326	U...G	0.163,0.129,,...,1.0	0.0, 0.137,...,0.0
4	0	0.542	G...C	1.0,0.932,...,0.5	0.0,0.072,...,0.0

5.2 High Attention Regions Calculation

PrismNet calculates a high attention region in each binding site to indicate the most important regions for RBP binding. Each binding site with length of 101 nt is split into sliding windows (window size: 20 nt, window step: 1 nt) and the average of both sequence and structure saliency values are calculated. The sliding window with the highest saliency value is defined as the high attention region. To calculate high attention regions, we provide the script *exp/prismnet/har.sh*, which can be run using:

```
$ exp/prismnet/har.sh TIA1_Hela /path/to/inference_file.tsv
```

The *TIA1_Hela* is the name of the trained model that you replace with the name of the data file you want to use. */path/to/inference_file.tsv* is the file containing the samples for which high attention regions need to be calculated.

The command output file *TIA1_Hela.har* contains the high attention regions of binding sites (Table 3).

5.3 Plot Image of Saliency Map

PrismNet can plot the image of a saliency map for visualization, which could be done by running the following script *exp/prismnet/saliencyimg.sh*.

Table 3
The data format for the high attention region file

Sample id	Binding probability	Start coordinate	End coordinate
0	0.066	88	108
1	0.212	68	88
2	0.331	71	91
3	0.461	37	57
4	0.157	17	37

```
$ exp/prismnet/saliencyimg.sh TIA1_Hela /path/to/inference_-
file.tsv
```

TIA1_Hela is the model name, */path/to/inference_file.tsv* is the file containing the samples for saliency map plotting, and the directory *TIA1_Hela_img* contains the images of saliency maps (Fig. 1).

6 Motif Construction

To characterize the sequence and structure binding preference of an RBP, we calculate the sequence-and-structure integrative motif, which contains a sequence and a structure component. The sequence component is like a conventional motif containing the positional-probabilities of four nucleotides (A, C, G, U) and the structural component containing the probability of paired or unpaired structural status of each nucleotide ("P" for paired, represents double-stranded RNA structure; "U" for unpaired, represents single-stranded RNA structure, Fig. 2).

To construct the sequence and structure integrative motifs from the trained PrismNet model, users can use the analysis code in the *motif_construct* directory.

Users should first install the tools RNAfold [20] and TOMTOM [21]. RNAfold is a tool to predict RNA secondary structure based on RNA sequence data and structure information (here icSHAPE score). TOMTOM is a tool to compare the motif similarity. Then, users can call the integrative motif from the saliency map file of the PrismNet model using:

```
$ perl saliency_motif.pl infile.sal outfile
```

The *infile.sal* is the saliency map file output from PrismNet model and *outfile* is the prefix of the output file.

This command line generates multiple output files that contain the motif information:

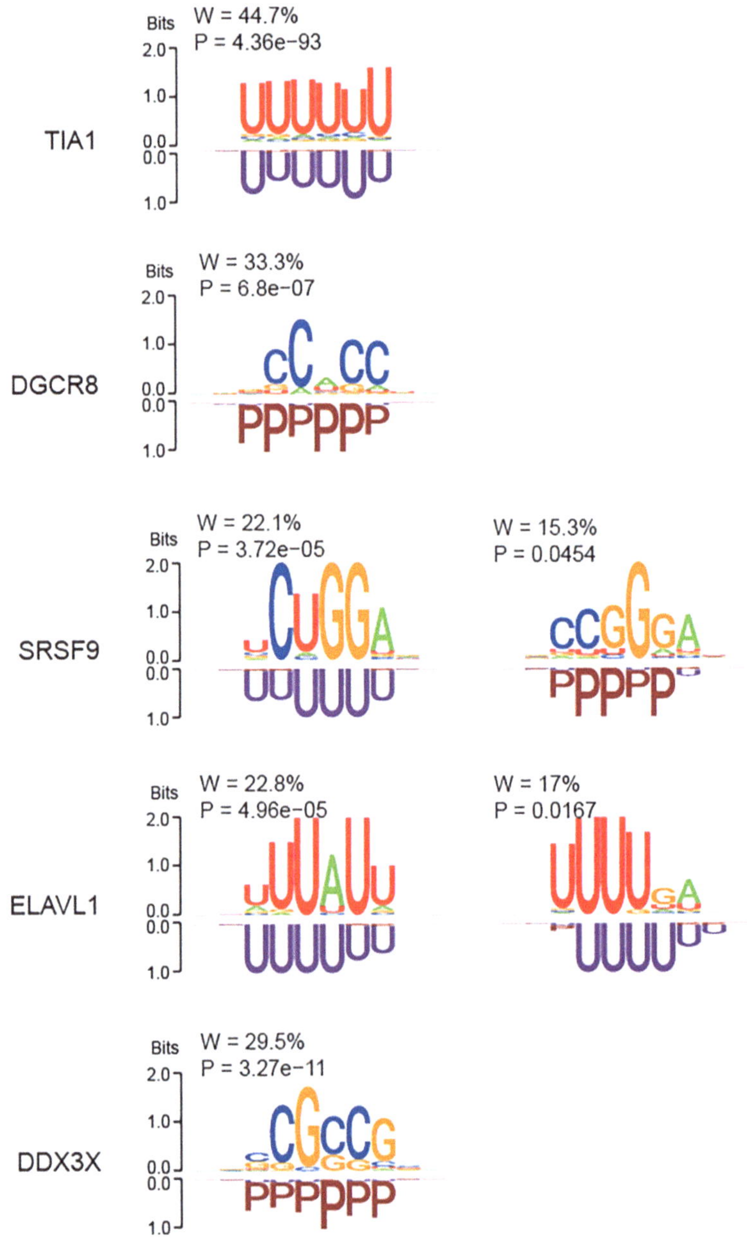

Fig. 2 Integrative motifs derived by PrismNet models. Each integrative motif contains both a sequence component (upper, "AUCG" represent four kinds of nucleotides) and a structure component (bottom, "P" represents double-stranded structure (paired) and "U" represents single-stranded structure (unpaired)). The motif weight (W) and *P* values are shown

The motif summary file *outfile_motif_summary.txt* contains the integrative motif and the number of motif sites and weights for the top 10 ranking motifs (Table 4).

Table 4
The file format of the motif summary file

Motif	Motif site	Motif weight
GGGGGC|PPPPPP	40	0.180995
GUGGUG|UUUUUU	36	0.162896
GGUAGC|UUUUUP	28	0.126697
UUUUUU|UUUUUU	28	0.126697
GGGUCG|PUUUUU	24	0.108597

The motif probability file *outfile_motif_prob.txt* contains the sequence and structure position weight matrix (PWM) of the top 10 ranking integrative motifs. The sequence PWM is a 4 × 6 matrix (p_{ij}, where $i = 1, 2, 3, 4$ corresponding to A, C, G, U, and j is the site index, $1 \leq j \leq 6$) and the structure PWM is a 2 × 6 matrix (p_{rj}, where $r = 1, 2$ corresponding to U, P, and j is the site index, $1 \leq j \leq 6$).

$$\sum_{i=1}^{4} p_{ij} = 1$$

$$\sum_{r=1}^{2} p_{rj} = 1$$

To evaluate motif enrichment, users can calculate the significance (P value) for each RBP motif, using:

```
$ Rscript motif_sig.R outfile_motif_summary.txt outfile_mo-
tif_sig.txt
```

By default, motifs with P values <0.05 are defined as significantly enriched integrative motifs (Fig. 2).

Acknowledgements

This work was supported by the State Key Research Development Program of China (Grant No. 2019YFA0110002 and the National Natural Science Foundation of China (Grants No. 32125007 and 91940306) and to Q.C.Z.

References

1. Gerstberger S, Hafner M, Tuschl T (2014) A census of human RNA-binding proteins. Nat Rev Genet 15:829–845
2. Licatalosi DD, Darnell RB (2010) RNA processing and its regulation: global insights into biological networks. Nat Rev Genet 11:75–87
3. Castello A, Fischer B, Eichelbaum K et al (2012) Insights into RNA biology from an atlas of mammalian mRNA-binding proteins. Cell 149:1393–1406
4. Baltz Alexander G, Munschauer M, Schwanhäusser B et al (2012) The mRNA-bound proteome and its global occupancy profile on protein-coding transcripts. Mol Cell 46: 674–690
5. Lambert N, Robertson A, Jangi M et al (2014) RNA Bind-n-Seq: quantitative assessment of the sequence and structural binding specificity of RNA binding proteins. Mol Cell 54: 887–900
6. Ellington AD, Szostak JW (1990) *In vitro* selection of RNA molecules that bind specific ligands. Nature 346:818–822
7. Ray D, Kazan H, Chan ET et al (2009) Rapid and systematic analysis of the RNA recognition specificities of RNA-binding proteins. Nat Biotechnol 27:667–670
8. Gilbert C, Svejstrup JQ (2006) RNA immunoprecipitation for determining RNA-protein associations in vivo. Curr Protocols Mol Biol Chapter 27:Unit 27.24
9. Licatalosi DD, Mele A, Fak JJ et al (2008) HITS-CLIP yields genome-wide insights into brain alternative RNA processing. Nature 456: 464–469
10. Hafner M, Landthaler M, Burger L et al (2010) Transcriptome-wide identification of RNA-binding protein and microRNA target sites by PAR-CLIP. Cell 141:129–141
11. Van Nostrand EL, Pratt GA, Shishkin AA et al (2016) Robust transcriptome-wide discovery of RNA-binding protein binding sites with enhanced CLIP (eCLIP). Nat Methods 13: 508–514
12. Bailey TL, Boden M, Buske FA et al (2009) MEME SUITE: tools for motif discovery and searching. Nucleic Acids Res 37:W202–W208
13. Maticzka D, Lange SJ, Costa F et al (2014) GraphProt: modeling binding preferences of RNA-binding proteins. Genome Biol 15:R17
14. Alipanahi B, Delong A, Weirauch MT et al (2015) Predicting the sequence specificities of DNA- and RNA-binding proteins by deep learning. Nat Biotechnol 33:831–838
15. Orenstein Y, Wang Y, Berger B (2016) RCK: accurate and efficient inference of sequence- and structure-based protein–RNA binding models from RNAcompete data. Bioinformatics 32:i351–i359
16. Sun L, Xu K, Huang W et al (2021) Predicting dynamic cellular protein–RNA interactions by deep learning using *in vivo* RNA structures. Cell Res 31:495–516
17. Jankowsky E, Harris ME (2015) Specificity and nonspecificity in RNA-protein interactions. Nat Rev Mol Cell Biol 16:533–544
18. Lewis CJT, Pan T, Kalsotra A (2017) RNA modifications and structures cooperate to guide RNA–protein interactions. Nat Rev Mol Cell Biol 18:202–210
19. Shi B, Zhang J, Heng J et al (2020) RNA structural dynamics regulate early embryogenesis through controlling transcriptome fate and function. Genome Biol 21:120
20. Lorenz R, Bernhart SH, Höner Zu Siederdissen C et al (2011) ViennaRNA package 2.0. Algorithms Mol Biol 6:26
21. Gupta S, Stamatoyannopoulos JA, Bailey TL et al (2007) Quantifying similarity between motifs. Genome Biol 8:R24

Chapter 9

High-Resolution Atomic Force Microscopy Imaging of RNA Molecules in Solution

Jienyu Ding

Abstract

Atomic force microscopy (AFM) is an important and versatile technique to investigate the structures and dynamics of biomolecules under physiologically relevant conditions at the single-molecule level. Recent progresses in high-resolution AFM imaging of nucleic acids have expanded this technique from simple characterization of double-stranded DNA or RNA to detailed analyses of the structure and dynamics of large functional RNAs with complex folds. Several technical developments, such as sharper probes and more stable instruments with novel imaging modes, AFM is capable of directly visualizing RNA conformational heterogeneity in solution in real time. Here, we introduce a comprehensive method for recording high-resolution images of RNA molecules, including sample preparation, instrument setup, data acquisition, and image processing.

Key words Solution atomic force microscopy, RNA conformational heterogeneity, High-resolution imaging methods

1 Introduction

Since its invention in 1986 [1], atomic force microscopy (AFM) has been recognized with great potential for studying biological samples. It measures very weak force interactions between the AFM probe and a sample's atomic surface, making it possible to construct three-dimensional topological maps of individual molecules immobilized on a solid support. Over the past 25 years, AFM has been applied extensively to study structure and dynamics of biological molecules, including nucleic acids [2–12], nucleic acid assemblies [13], and nucleic acid–protein complexes [14–19]. High-resolution probes (1–10 nm) can achieve detailed images of molecular surfaces under a variety of solution conditions and experimental settings. Such surface information is used to characterize the structure and dynamics of nucleic acids, proteins, and nucleic acid–protein complexes. Continued developments in

Jienyu Ding et al. (eds.), *RNA Structure and Dynamics*, Methods in Molecular Biology, vol. 2568,
https://doi.org/10.1007/978-1-0716-2687-0_9,

instrument hardware, probe sensitivity, and software for data acquisition and image processing render AFM as an increasingly powerful tool to visualize and probe biomolecules at single-molecule resolution [20–30]. Compared to other structural biology techniques (*e.g.*, X-ray crystallography, NMR, cryo-EM), AFM consumes very little sample, requires no special modification of treatment, such as crystallization, freezing, staining, or labeling, and can measure under physiologically relevant solution conditions. As such, high-resolution AFM in solution opens an entire novel and attractive area for structural biology.

Although tremendous progress has been made in the imaging of nucleic acids by AFM, the method's spatial resolution limitation has hampered its potential for biological applications [31]. Recently, the Moreno-Herrero group achieved an AFM resolution of ~3 nm for double-stranded RNA [2], revealing discernable major and minor grooves. Even more recently, the Scheuring group developed a post-acquisition localized image reconstruction method, localization AFM (LAFM), capable of resolving single amino acid residues with sub-nanometer resolution [27]. These methods might useful for extracting local higher-resolution information from AFM data, but may be less effective for highly dynamic structures like folded RNAs. In either case, obtaining the highest quality AFM images is necessary for maximizing the attainability of structural information therein.

The requirements for high-resolution imaging of RNAs are described [32], the most critical of which includes the use of a sharp probe and gentle tapping mode. Although some imaging artifacts can be dealt with at the processing stage, they should be avoided whenever possible to mitigate negative effects on image quality. Such artifacts can occur for a variety of reasons. These include internal factors, such as scanner effects (piezo creep and edge overshoot), mechanical vibration, thermal drift, and light interference, as well as external factors, such as acoustic noise, mechanical or building vibrations, air currents, electronic noise, temperature waves, and image processing [33–36]. In this chapter, we describe a detailed protocol for obtaining high-resolution images of RNA molecules under physiologically relevant solution conditions using AC mode (also known as tapping mode) (Fig. 1) as well as useful tips for reducing or minimizing imaging artifacts.

2 Materials

1. AFM probes designed for imaging soft biological samples (*see* **Note 1**), *e.g.*, Biolever mini BL-AC40TS cantilevers (Oxford Instruments, Abingdon, United Kingdom), which have a nominal radius of 8 nm, a resonance frequency of ~110 kHz, and a nominal spring constant of 0.09 N/m. Such probes can

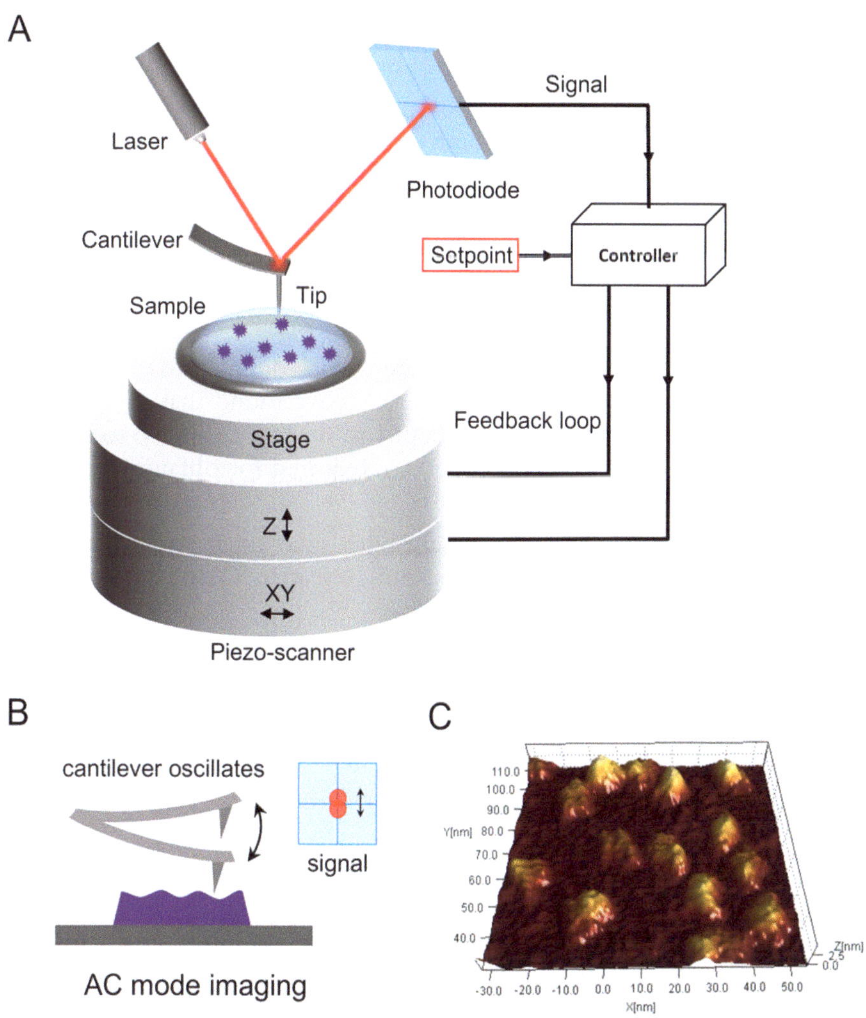

Fig. 1 Imaging RNA samples in solution using AFM. (**a**) Diagram of instrument setup. A nanoscale tip is attached to a cantilever that forms a spring. As the tip probes the molecular surface, the tip-sample interaction force is detected by a laser diode and a split photodetector. The feedback loop continuously monitors the interaction force and maintains a constant deflection through the piezo-scanner. (**b**) AC mode imaging. In AC mode, the cantilever is caused to vibrate near its resonance frequency by a small piezo electric actuator very near the cantilever. The oscillating movement of the tip is described as a sinusoidal motion. When the tip comes near the sample, this motion is reduced by attractive or repulsive interactions. The detector senses the oscillatory motion of the cantilever and the electronics inside the controller measure the amplitude and phase of this oscillation with respect to the drive signal. The feedback loop keeps the amplitude of this oscillatory motion constant by adjusting the XYZ position of the piezo-scanner. (**c**) A representative 3D AFM image

occasionally give high-resolution images but have poor reproducibility.

2. Ultrasharp, high-resolution probes, *e.g.*, FASTSCAN-D-SS cantilevers (Bruker), with a nominal tip radius of ~1 nm, a resonance frequency of 80–140 kHz, and a nominal spring constant of ~0.25 N/m.

3. Muscovite mica (Grade V1) (Ted Pella Redding, CA).
4. 15 mm AFM/STM metal specimen discs
5. Fast-setting epoxy, *e.g.*, DOUBLEBUBBLE® (Hardman, Ellsworth Adhesives, Germantown, WI).
6. 1-(3-Aminopropyl) silatrane (APS): not commercially available, but can be synthesized according to the protocols reported elsewhere [17, 37].
7. RNA sample (~10–20 nm): Here we use *Tte* adenosylcobalamin riboswitch aptamer domain.
8. Low-salt buffer: Here we use 0.5 mM MES, pH 6.0, 10 mM KCl, 1 mM $MgCl_2$. We have found that low-salt conditions, *e.g.*, 10 mM KCl and 1 mM Mg^{2+} (if shown to help with RNA fold and stability), are required for obtaining high-resolution images.
9. High-resolution atomic force microscope. Here we use a Cypher VRS (Asylum Research, Oxford Instruments, Santa Barbara, CA).
10. Image processing software: SPIP/MountainsSPIP® (Scanning Probe Image Processor, commercially available) or Gwyddion (version 2.56, free).
11. Ultrapure deionized water.
12. Filtered nitrogen gas.
13. Scotch® tape.

3 Methods

3.1 Sample Preparation

3.1.1 Mica Substrate Preparation

Various substrates (*e.g.*, mica, glass, highly oriented pyrolytic graphite (HOPG), gold films) have been used for imaging biological samples [10]. The smoothness of mica surface, however, makes it the most widely used substrate for biomolecules, especially nucleic acids. Mica discs (ready-to-use or punched from mica sheets) are pre-fixed on 15 mm AFM/STM metal supports using epoxy (Fig. 2, Step I). Before each use, the mica surface is cleaved by stripping the surface with Scotch® tape, leaving uncompensated negative charges on the surfaces. For nucleic acids or other negatively charged specimens, the mica surface first needs to be treated either with divalent cations or by chemical modification.

3.1.2 Chemical Modification of Mica with APS

APS-mica is an improvement over AP-mica (3-aminopropyltriehoxy silane (APTES)-treated mica), which is resistant to hydrolysis during time-lapse AFM imaging [37]. Compared to the divalent cation (Mg^{2+}, Ni^{2+}, Zn^{2+}, et al.) modification method, APS-mica tolerates a broader range of buffers and pH (up to 10). For RNA, it also avoids potential structural perturbations caused by divalent

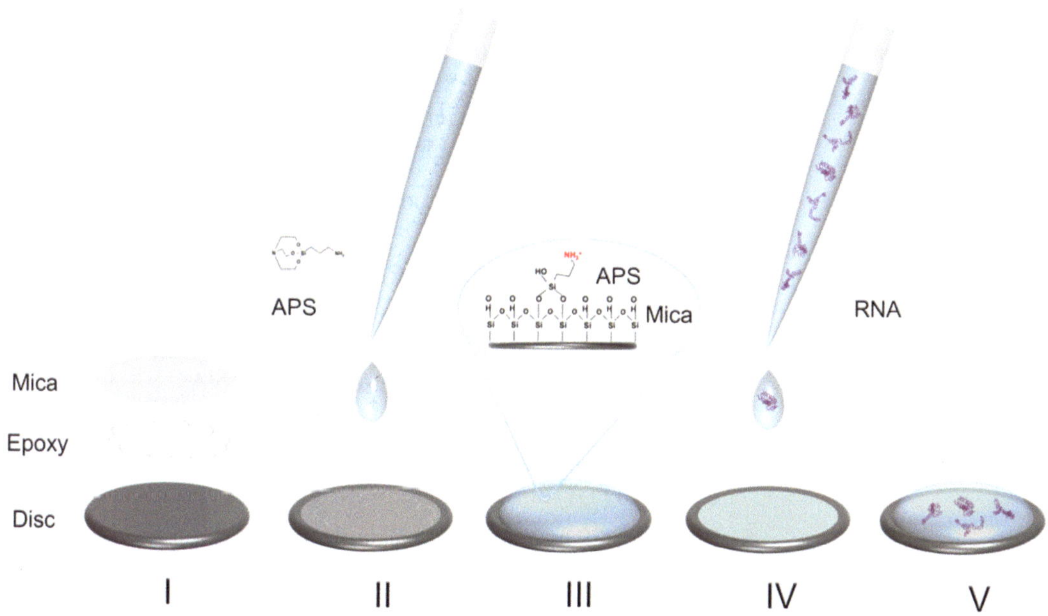

Fig. 2 RNA sample preparation for AFM. Step I: mica substrate preparation. The mica disc is fixed on the metal support using epoxy. Steps II and III: coat mica with APS. The mica surface is cleaved using Scotch® tape. Deposit 100 μl of APS (~0.17 mM) onto the freshly cleaved mica and wait at least 30 min. Rinse the mica surface with ultrapure deionized water and dry with filtered nitrogen gas. Steps IV and V: immobilize RNA sample on the APS-mica surface. Deposit 10 μl 10–20 nM RNA sample on the APS-mica and wait for 10–20 min. Rinse the surface with one or more aliquots of 200 μl buffer by gentle pipetting to remove unbound molecules. Add 100 μl AFM buffer for imaging

cations. Moreover, the uniform and smooth treated mica surface allows for reliable imaging of nucleic acids in solution and the low surface-charge density indicates that the geometry of immobilized nucleic acids can resemble the 2D projection of unperturbed molecules onto a plane, suggesting that this immobilization method is suitable to study global RNA conformations. The following steps outline the basic procedure for preparing APS-treated mica (*see* **Note 2**) for high-resolution imaging of RNA (Fig. 2, steps II and III).

1. Prepare 50 mM APS in ultrapure deionized water and store it at 4 °C.
2. Dilute 300-fold just before use.
3. Cleave mica with Scotch® tape.
4. Place 100 μl diluted APS on the freshly cleaved mica surface.
5. Wait at least 30 min.
6. Rinse the mica surface with running ultrapure deionized water to remove excess APS.
7. Dry the mica surface using a gentle stream of filtered nitrogen gas.

3.1.3 RNA Sample Preparation for High-Resolution AFM

The RNA sample used for high-resolution imaging is in vitro transcribed using linear DNA templates, and purified by native gel electrophoresis. For AFM studies, a 1 ml transcription reaction volume is more than sufficient in most cases. The whole sample preparation can be done within two days.

1. Prepare a 1 ml in vitro transcription reaction and incubate at 37 °C for 3 ~ 5 h.
2. Load the 1 ml reaction to two one-well native polyacrylamide gels (0.5 ml per gel).
3. Run gels at room temperature at 200 V for an appropriate amount of time, based on the size of the RNA.
4. Excise the desired transcript, and elute the RNA from the gel using gel elution buffer (*e.g.*, 50 mM sodium acetate, pH 5.3, 2 mM EDTA) and incubate overnight with agitation at 4 °C.
5. Buffer exchange the RNA sample to AFM buffer (*see* **Note 3**) using a centrifugal filter with the appropriate molecular weight cut-off (*see* **Note 4**).
6. Place 10 μl of RNA (10–20 nM) on freshly prepared APS-mica (Fig. 2, steps IV and V).
7. Wait 10–20 min.
8. Wash mica surface gently with one or more aliquots of 200 μl AFM buffer to remove unbound or weakly associated molecules (*see* **Note 5**).
9. Add 100 μl AFM buffer for imaging (*see* **Note 6**).

3.2 Atomic Force Microscopy

3.2.1 AFM Setup

Different AFM instruments use different control software, imaging modes, setup protocols, and procedures. Here we describe a protocol for RNA imaging in solution using the Cypher VRS AFM (Asylum Research) in AC mode (also known as tapping mode) (*see* **Note 7**). The Cypher VRS AFM is specifically designed for high-speed AFM, but we found that it is also very good for high-resolution imaging in solution.

1. Turn on Asylum Research software (Igor Pro) and select AC Molecule mode under the Bio tab in the Mode Master (*see* **Note 8**).
2. Initialize motors. This will take several minutes.
3. Open the environment panel and set the target temperature to 4 °C. For high-resolution imaging, low temperature is preferred to minimize Brownian motion of RNA molecules on the mica surface.
4. Open the valve to allow filtered nitrogen gas flow into the chamber to keep the cantilever holder and other components inside the chamber dry. This is especially important at 4 °C to minimize condensation.

5. Calibrate the cantilever in air (For FASTSCAN-D-SS cantilever, this step is skipped).
   (a) Click "Focus on tip" and use "Move Focus" to set the tip Z position.
   (b) Click "Focus on the sample" and use "Move Focus" to set the Z position for the sample surface.
   (c) Click "SpotOn" then click on the cantilever. The system will automatically position the cantilever at this spot on.
   (d) Adjust the laser spot position manually to get the maximum "Sum" value.
   (e) Click "Zero PD" to center the spot on the photodiode and zero out the deflection signal.
   (f) Open the Thermal Graph panel by clicking "Thermal" in the Master Panel.
   (g) Calibrate the cantilever by clicking "GetReal" in the Thermal Graph panel. The sensitivity ("Amp InvOLS" parameter) and spring constant of the cantilever are calculated from the thermally driven resonance peak. The spring constant is a property of the cantilever and will not change during imaging. The sensitivity will change with the optical path, which depends on the laser position, imaging condition (air or solution), and cantilever position.
6. Tune the cantilever in solution.
   (a) Place metal support containing RNA sample immobilized on APS-mica onto the magnetic stage inside of the AFM chamber.
   (b) Gently fill cantilever holder tubing with buffer using syringes, creating a small droplet around the cantilever.
   (c) Manually approach the sample to the cantilever by turning the control knob to raise the AFM stage, leaving ~2–3 mm liquid between the cantilever and sample surface.
   (d) Repeat tip-calibration steps a-f in **step 5**.
   (e) Click "Move to Pre-Engage" to move the tip Z position to 50 μm above the defined sample surface. The laser spot may move on the cantilever during this step. Reposition the laser to maximize Sum value and zero the Deflection if the laser spot moves.
   (f) Click "Capture" in the Thermal Graph panel to collect a thermal profile in solution.
   (g) Zoom on the first peak and click "Fit" to fit the thermal peak. The sensitivity will be updated.
   (h) Turn on BlueDrive and position the blue laser on the back (furthest from the tip) of the cantilever.

(i) Adjust blue laser position manually to maximize the amplitude signal.

(j) Set up Auto Tune parameters in the Tune Parms panel. Set the low and high parameters to bracket the drive frequency range of the cantilever. For Biolever mini BL-AC40TS, use 10 kHz and 50 kHz, respectively.

(k) Set the Target Amplitude to 500 mV.

(l) Click "Auto Tune" in the Cantilever Tune panel to automatically identify the resonance peak and select a Drive Amplitude.

7. Set up initial parameters for tip approach. For sharp tips, small scan size and a low scan rate are good to protect the tip. As a rule of thumb, good starting values are as follows: Scan Size: 0.5 μm, Points & Lines: 512 × 256, Scan Rate: 1 Hz, Setpoint: 450 mV, Integral Gain: 30. The setpoint will be adjusted during the tip approach.

8. Tip approach. It is recommended to first set Engage Approach Speed to "Slower" to avoid collision (*see* **Note 9**). Click "Start Tip Approach" to engage. Reduce the setpoint value by a small amount (*e.g.*, 20 mV), and Start Tip Approach again. Repeat the process iteratively until the tip is just above the mica surface. This can be assessed quickly by acquiring an image by clicking "Frame Up" or "Frame Down."

3.2.2 Image Acquisition

1. Set the acquisition parameters in the Master Panel as follows: Scan Size: 0.5 μm, Points & Lines: 1024 × 1024, Scan Rate: 1 Hz (*see* **Note 10**).

2. Name your file in the "Base Name" in the "Save Options" and put "Note" if you have.

3. After all the parameters are set, click "Frame Up" or "Frame Down" to start collecting. "Setpoint" and "Integral Gain" can and should be adjusted during the imaging process, depending on the imaging stability. It will take around 18 min to collect one image using the above settings. The imaging resolution should be around 5 Å. The imaging will continue until stopped, and each completed image will be automatically saved as *.ibw files in the folder created by the software in the very beginning.

3.2.3 Image Processing

The AFM images can be processed and analyzed by imaging software that can read the .ibw file format. Gwyddion is the most popular free AFM-image processing software. However, we recommend the commercial software SPIP™ (Scanning Probe Image Processor, updated to MountainsSPIP in 2021) for more advanced processing. Regardless of which software is used, the basic principles for image processing are the same. Here we use SPIP™ to illustrate the procedure for image processing.

1. Open the .ibw file. Click "File → open/browse," and choose the file you want to open. Usually, we use the "height" file for analysis and image processing.
2. Select the particles based on the height scale bar on the right of the image. The highlighted areas will be masked out and excluded during the plane correction (*see* **Note 11**).
3. Plane correction. Plane correction is the first and most important step in AFM image processing. The purpose of plane correction is to remove the effects of tilted background or scanner bow, both of which are common problems in AFM images. This can be done by leveling the AFM images. A widely used method for leveling is polynomial fitting. In SPIP™, this can be done using the built-in settings by clicking "Modify → Quick Launch → Profile Leveling." In many cases, third Order Polynomial fitting is sufficient.
4. Filtering. Filtering is used to remove the high- or low-frequency noise from the AFM images. Matrix filtering and Fourier transform-based filtering are most commonly used.
   (a) Matrix filters are typically grouped into low-pass and high-pass filters. Low-pass filters allow only low-frequency signals, resulting in the smoothing of the image. However, the overuse of low-pass filters will blur the image. Similarly, high-pass filters allow only high-frequency signals, resulting in the sharpening of the image. Overuse of high-pass filters will result in image artifacts. In SPIP™, there are many other matrix filters for varying purposes, such as different kinds of noise reduction, smoothing, and roughness filters. Users can access these filters by clicking "Modify → Quick launch in Filtering."
   (b) Fourier transform-based filtering relies on transforming an image from real space to frequency space, removing some specific components, and then back to real space. It is much more flexible than matrix filtering and can remove or enhance certain components of the image. Users can find this function by clicking "Analyze → FFT/PSD analysis."
5. Error correction. Errors happen often during AFM imaging. Most such imaging artifacts are due to the unwanted interactions between sample and tip, such as sample drift, sample sticking to the tip, or vibration/streaking patterns from using too much applied force. Other errors can result from external vibrations or acoustic noise (*see* **Note 12**). To remove artifacts during image processing, one can apply the "line removal" tool. In SPIP™, this function is accessed as "Modify → Quick Launch in Filtering → De-spiking." Another well-

known type of error is tip dilation, also known as tip-sample convolution. This can be removed by "tip deconvolution," incorporated in most AFM image processing software when the tip shape is well known (either by blind tip estimation or measurement). In SPIP^{TM}, this is done through the function "Tip Characterization," accessed under "Analyze → Tip" (*see* **Note 13**).

6. Image display. How an image is displayed is critical to its interpretation. To show the features most clearly, one needs to select an appropriate height range and color to display the image. AFM images are usually displayed in a color image with a color scale bar on the right, which represents height. The scale bar may have a very broad range that concentrates the data in a certain narrow range because of the small number of outlying data points. These outlying data points may come from the particle topological features. But in most cases, they are from unwanted errors or imaging artifacts. As a result, the image shows few details and low contrast (*see* **Note 14**).
   (a) Height range. To enhance the visibility of certain features, one can adjust the scale bar by adjusting the lower and upper height values for the display. This will not change the data but redistribute the color over the given height range. The data points with height values lower than the set range will appear dark, while data points with height values greater than the set range will appear bright. In SPIP™, users can adjust the color range in the height histogram by clicking "Analyze → Histogram."
   (b) Color selection. The color chosen to display the AFM images is very subjective. However, some colors may help with visualizing certain features in the image. In general, using a palette with more than one color gradient gives high contrast for an image with various heights. Color schemes can be selected by clicking "General → 2D Image Color Scale → Built-in Color Scales."
7. Save the processed file. Usually, the software gives multiple options for saving the processed image as different file-types.

4 Notes

1. To minimize tip-related imaging artifacts, the choice of an optimal probe for a given experiment is important: the smaller the object size, the sharper the tip required. For soft samples, such as biomolecules, the softness of the tip is also critical to reduce molecular deformations.

2. In practice, APS-mica can be kept for ~1-week, but freshly prepared APS-mica just prior to imaging is highly recommended.
3. AFM buffer should be ultrafiltered (0.22 μm), free of precipitates or any solid particulates. This is very important for obtaining a clean image free of artifacts.
4. If you do not plan to image the sample immediately, stop here and store the sample in whatever conditions that favor the sample.
5. The sample incubation time (**step 7**) and the extent of washing can be varied to optimize particle density for imaging.
6. Once a sample is loaded on APS-mica, it needs to be imaged immediately.
7. The cantilever calibration and tuning steps outlined in this section are for Asylum cantilevers, which are compatible with Asylum AFM instruments. For non-Asylum cantilevers, such as FASTSCAN-D-SS, the Auto Tune function may not work properly. In this situation, the resonance frequency can first be approximated from the thermal peak, and then adjusted using the Tune function and manually selecting the resonance peak.
8. The AFM should be turned on at least 1 h prior to imaging to allow ample time for the instrument to warm up and stabilize. This will help with imaging stability.
9. Tip-sample approach is a critical step in the whole AFM setup, as it is easy to crash the tip during this step, especially sharp tips. Be patient and use a slow approach to land on the sample.
10. The scan speed may need to be reduced if edge overshoot artifacts occur.
11. To obtain accurate sizes of particles from AFM images, the instrument's XYZ image dimensions need to be well calibrated.
12. To avoid external mechanical vibrations from the building, such as floor vibrations and acoustic vibrations, it is recommended to install the instrument on a vibration-free table near a corner instead of at the center of a room, and on the lowest floor (if possible). It may also be advantageous to carry out experiments during the night or weekend when electronic and mechanical noise may be minimal.
13. Be cautious with error correction during image processing since it may alter the data and bias analysis and interpretation.
14. Be cautious with color-scale adjustment. Some image processing software may discard (as opposed to exclude) the data outside the set range when histogram adjustment is performed, resulting in a loss of data.

Acknowledgments

This research was supported by the Intramural Research Program of the NIH, National Cancer Institute, Center for Cancer Research (Y-X. Wang).

References

1. Binnig G, Quate CF, Gerber C (1986) Atomic force microscope. Phys Rev Lett 56:930–933
2. Ares P, Fuentes-Perez ME, Herrero-Galan E et al (2016) High resolution atomic force microscopy of double-stranded RNA. Nanoscale 8:11818–11826
3. Kalle WH, Macville MV, Van De Corput MP et al (1996) Imaging of RNA in situ hybridization by atomic force microscopy. J Microsc 182:192–199
4. Kuznetsov YG, Daijogo S, Zhou J et al (2005) Atomic force microscopy analysis of icosahedral virus RNA. J Mol Biol 347:41–52
5. Kuznetsov YG, Dowell JJ, Gavira JA et al (2010) Biophysical and atomic force microscopy characterization of the RNA from satellite tobacco mosaic virus. Nucleic Acids Res 38: 8284–8294
6. Kuznetsov YG, Mcpherson A (2006) Atomic force microscopy investigation of Turnip Yellow Mosaic Virus capsid disruption and RNA extrusion. Virology 352:329–337
7. Lymans'kyi OP (2007) Visualization of RNA transcripts with atomic force microscopy. Tsitol Genet 41:12–18
8. Lyubchenko YL, Gall AA, Shlyakhtenko LS et al (1992) Atomic force microscopy imaging of double stranded DNA and RNA. J Biomol Struct Dyn 10:589–606
9. Lyubchenko YL, Shlyakhtenko LS (1997) Visualization of supercoiled DNA with atomic force microscopy in situ. Proc Natl Acad Sci USA 94:496–501
10. Lyubchenko YL, Shlyakhtenko LS, Ando T (2011) Imaging of nucleic acids with atomic force microscopy. Methods 54:274–283
11. Ng JD, Kuznetsov YG, Malkin AJ et al (1997) Visualization of RNA crystal growth by atomic force microscopy. Nucleic Acids Res 25:2582–2588
12. Pino S, Biasiucci M, Scardamaglia M et al (2011) Nonenzymatic ligation of an RNA oligonucleotide analyzed by atomic force microscopy. J Phys Chem B 115:6296–6303
13. Hansma HG, Oroudjev E, Baudrey S et al (2003) TectoRNA and 'kissing-loop' RNA: atomic force microscopy of self-assembling RNA structures. J Microsc 212:273–279
14. Henn A, Medalia O, Shi SP et al (2001) Visualization of unwinding activity of duplex RNA by DbpA, a DEAD box helicase, at single-molecule resolution by atomic force microscopy. Proc Natl Acad Sci USA 98:5007–5012
15. Kiselyova OI, Yaminsky IV, Karger EM et al (2001) Visualization by atomic force microscopy of tobacco mosaic virus movement protein-RNA complexes formed *in vitro*. J Gen Virol 82:1503–1508
16. Limanskaia O, Limanskii AP (2008) Visualization of elongation complexes for t7 Rna polymerase by atomic force microscopy. Mol Biol (Mosk) 42:533–542
17. Lyubchenko YL, Shlyakhtenko LS (2009) AFM for analysis of structure and dynamics of DNA and protein-DNA complexes. Methods 47:206–213
18. Smith BL, Gallie DR, Le H et al (1997) Visualization of poly(A)-binding protein complex formation with poly(A) RNA using atomic force microscopy. J Struct Biol 119:109–117
19. Endo M, Tatsumi K, Terushima K et al (2012) Direct visualization of the movement of a single T7 RNA polymerase and transcription on a DNA nanostructure. Angew Chem Int Ed Engl 51:8778–8782
20. Kueng A, Kranz C, Lugstein A et al (2003) Integrated AFM-SECM in tapping mode: simultaneous topographical and electrochemical imaging of enzyme activity. Angew Chem Int Ed Engl 42:3238–3240
21. Trohalaki S (2012) Multifrequency force microscopy improves sensitivity and resolution over conventional AFM. MRS Bull 37:545–546
22. Kocun M, Labuda A, Meinhold W et al (2017) Fast, high resolution, and wide modulus range nanomechanical mapping with bimodal tapping mode. ACS Nano 11:10097–10105
23. Sokolov I, Dokukin ME (2018) Imaging of soft and biological samples using AFM ringing mode. Methods Mol Biol 1814:469–482

24. Dzedzickis A, Bucinskas V, Virzonis D et al (2018) Modification of the AFM sensor by a precisely regulated air stream to increase imaging speed and accuracy in the contact mode. Sensors (Basel) 18:2694
25. Koehler M, Fis A, Gruber HJ et al (2019) AFM-based force spectroscopy guided by recognition imaging: a new mode for mapping and studying interaction sites at low lateral density. Methods Protoc 2:6
26. Kunicki P, Angelov T, Ivanov T et al (2019) Sensitivity improvement to active piezoresistive afm probes using focused ion beam processing. Sensors (Basel) 19:4429
27. Heath GR, Kots E, Robertson JL et al (2021) Localization atomic force microscopy. Nature 594:385–390
28. Dufrene YF, Ando T, Garcia R et al (2017) Imaging modes of atomic force microscopy for application in molecular and cell biology. Nat Nanotechnol 12:295–307
29. Dufrene YF, Martinez-Martin D, Medalsy I et al (2013) Multiparametric imaging of biological systems by force-distance curve-based AFM. Nat Methods 10:847–854
30. Garcia R, Proksch R (2013) Nanomechanical mapping of soft matter by bimodal force microscopy. Eur Polym J 49:1897–1906
31. Heath GR, Scheuring S (2019) Advances in high-speed atomic force microscopy (HS-AFM) reveal dynamics of transmembrane channels and transporters. Curr Opin Struct Biol 57:93–102
32. Ares P, Gomez-Herrero J, Moreno-Herrero F (2018) High-resolution atomic force microscopy imaging of nucleic acids. Methods Mol Biol 1814:3–17
33. Ricci D, Braga PC (2004) Recognizing and avoiding artifacts in AFM imaging. Methods Mol Biol 242:25–37
34. Golek F, Mazur P, Ryszka Z et al (2014) AFM image artifacts. Appl Surf Sci 304:11–19
35. Canale C, Torre B, Ricci D et al (2011) Recognizing and avoiding artifacts in atomic force microscopy imaging. Methods Mol Biol 736:31–43
36. Voigtlander B (2019) Artifacts in AFM. In: Atomic Force Microscopy. Nanosci Technol. Springer, Cham. https://doi.org/10.1007/978-3-030-13654-3_8
37. Shlyakhtenko LS, Gall AA, Filonov A et al (2003) Silatrane-based surface chemistry for immobilization of DNA, protein-DNA complexes and other biological materials. Ultramicroscopy 97:279–287

Chapter 10

3D Structural Analysis of Long Noncoding RNA by Small Angle X-ray Scattering and Computational Modeling

Jie Zhang, Binxian Chen, and Xianyang Fang

Abstract

Small angle X-ray scattering (SAXS) has been widely applied as an enabling integrative technique for comprehensive analysis of the structure of biomacromolecules by multiple, complementary techniques in solution. SAXS in combination with computational modeling can be a powerful strategy bridging the secondary and 3D structural analysis of large RNAs, including the long noncoding RNAs (lncRNA). Here, we outline the major procedures and techniques in the combined use of SAXS and computational modeling for 3D structural characterization of a lncRNA, the subgenomic flaviviral RNA from Zika virus. lncRNA production and purification, RNA buffer and sample preparation for SAXS experiments, SAXS data collection and analysis, SAXS-aided RNA 3D structure prediction, and computational modeling are described.

Key words Small angle X-ray scattering, Computational modeling, Long noncoding RNA, RNA 3D structure, Subgenomic flavivirus RNA

1 Introduction

Long noncoding RNAs (lncRNAs) are an expanding group of cellular transcripts that range from 200 nucleotides (nts) to over 100 kb in length and possess little or no protein-coding potential [1]. Accumulating evidence have shown a myriad of functions for lncRNA in diverse cellular processes, including regulation of gene expression at almost every level [2]. Unsurprisingly, lncRNAs are crucially connected to diseases and can be potential therapeutic targets [3]. The diverse roles of lncRNAs are dictated by their propensities to form complex secondary and three-dimensional (3D) structures [4]. Currently, a variety of experimental methods such as SHAPE chemical probing and bioinformatics prediction have been developed to characterize the secondary structures of

Jienyu Ding et al. (eds.), *RNA Structure and Dynamics*, Methods in Molecular Biology, vol. 2568,
https://doi.org/10.1007/978-1-0716-2687-0_10,

lncRNAs in vitro or in vivo [5]. By contrast, experimental 3D structural characterization of lncRNAs remains challenging using the conventional high-resolution structural techniques including X-ray crystallography (XRC), nuclear magnetic resonance (NMR), and cryo-electron microscopy (cryo-EM), due to their large sizes and inherent flexibility [4, 6]. So far, only very few lncRNAs have been characterized at the 3D level [4, 7].

Over the last decades, major advances in instrument availability and data analysis strategies have led to exciting developments in the application of SAXS in structural biology including bridging the gap between the secondary and 3D structural analysis of large RNAs in solution [8, 9]. Although SAXS is limited with a moderate resolution, it remains one of the few techniques that can probe the structure and dynamics of macromolecules without size limitation under native conditions. SAXS data can provide information about the size, oligomeric state, and overall shape of the molecule. SAXS has been routinely utilized to validate the structural models, measure structural responses of RNA upon divalent cations and ligand binding, or reconstructing *ab initio* low-resolution 3D structures [8, 10–12]. Over the years, SAXS has emerged as an enabling integrative technique for comprehensive analysis of the structure and dynamics of biomacromolecules by multiple, complementary techniques in solution [8]. For example, SAXS can aid in NMR-driven RNA structure refinement and model building [13, 14]. As experimental determination of RNA structures remains time-consuming and technically challenging, a variety of RNA 3D structure prediction and modeling programs have been developed. However, "purely theoretical" 3D structural prediction of RNAs usually suffers from limited accuracy [15]. SAXS in combination with computational modeling can overcome such limitations [16], and has recently been employed to directly visualize the 3D solution structures of a group of lncRNAs, the subgenomic flaviviral RNAs (sfRNAs) from Dengue virus (DENV), Zika virus (ZIKV), and West Nile virus (WNV) [7], which are implicated to impact flavivirus replication, cytopathicity, and pathogenicity [17].

In this chapter, the procedures and techniques of the combined use of SAXS and computational modeling for 3D structural characterization of ZIKV sfRNA in solution are outlined. We describe how to produce and purify the lncRNA, prepare the buffer and lncRNA sample for SAXS measurements, collect and analyze SAXS data including *ab initio* shape reconstruction, obtain 3D structure models, and characterize the accessible 3D conformational ensemble of ZIKV sfRNA in solution by SAXS-aided computational modeling. Readers interested in the theoretical aspects of the methods are referred to reviews and guidebooks [18–20].

2 Materials

2.1 DNA Molecular Cloning

1. Plasmid vector encoding the full length lncRNA.
2. Desalted DNA oligonucleotides for subconstruct cloning (*see* **Note 1**).
3. PCR thermal cycler.
4. Enzymes and reagents including *Pfu* or *Taq* DNA polymerase and dNTP mix (A, C, G, T).

2.2 RNA Production by In Vitro Transcription

1. Desalted DNA oligonucleotide primers (*see* **Note 2**).
2. T7 RNA polymerase (*see* **Note 3**).
3. 10× transcription buffer: 400 mM Tris-HCl pH 8.0, 100 mM DTT, 20 mM spermidine.
4. 10× rNTP mix (40 mM of ATP, CTP, GTP, UTP).
5. RNase inhibitors.
6. Water bath with temperature control.
7. 5–10% denaturing polyacrylamide gel (acrylamide/bisacrylamide 19:1 ratio, 8 M urea, 1× TBE) or 5–10% native polyacrylamide gel (acrylamide/ bisacrylamide 19:1 ratio, 0.5 × TBM)
8. 10× TBE buffer: 1 M Tris, 1 M boric acid, 20 mM EDTA.
9. 10× TBM buffer: 1 M Tris, 1 M boric acid, 60 mM $MgCl_2$.
10. 2× RNA denaturing gel loading dye: 8 M urea, 0.05% bromophenol blue, and 0.05% xylene cyanol dissolved in 1× TE buffer, 0.5 M EDTA.
11. 2× RNA native gel loading dye: 3% Ficoll-400, 0.05% bromophenol blue, and 0.05% xylene cyanol dissolved in 1× TE buffer.

2.3 RNA Purification by Preparative Chromatography

1. Äkta pure HPLC system (GE Healthcare Life Sciences).
2. Superdex 75 increase 16/600 column (GE Healthcare Life Sciences).
3. Superdex 200 increase 16/600 column (GE Healthcare Life Sciences).
4. Size exclusion chromatography (SEC) buffer: 20 mM Tris-HCl pH 7.5, 100 mM KCl, 0.5 mM $MgCl_2$, 2 mM DTT, 2% glycerol.
5. HiTrap HP Q 5 ml anion exchange column (GE Healthcare Life Sciences).
6. Anion exchange buffer QA: 50 mM Tris-HCl pH 7.5, 50 mM KCl, 0.5 mM $MgCl_2$.
7. Anion exchange buffer QB: 50 mM Tris-HCl pH 7.5, 1 M KCl, 0.5 mM $MgCl_2$.

2.4 RNA Purification by Preparative Native PAGE

1. Power supply.
2. Electrophoresis chamber.
3. 22 × 22 cm^2 glass plates.
4. 3-mm-thick spacers.
5. 3-mm-thick combs.
6. High-vacuum grease (Dow Corning).
7. Binder clip 25 mm.
8. Plastic wrap.
9. TLC plate with fluorescent indicator 254 nm (Merck).
10. UV light source.
11. Scalpel blade.
12. 10 × Soak buffer: 200 mM Tris-HCl pH 7.5, 3 M NaOAc pH 5.2, 5 mM EDTA.

2.5 RNA Quantification with UV-Vis Spectroscopy

1. *NanoDrop* 2000 spectrophotometer.
2. OligoAnalyzer or Oligo Calc web server.

2.6 Sample and Buffer Preparation for SAXS

1. SpectraPor® dialysis tubing and membranes, or Slide-A-Lyzer™ dialysis cassettes.
2. HiTrap desalting column (GE Healthcare Life Sciences).
3. Amicon Ultra Centrifugal Filters (Merck) with adapted volume and molecular weight cut-off (MWCO).

2.7 SAXS Data Collection

1. Synchrotron-based SAXS instrument.
2. Sterile syringe filters 0.22 μM (Millex-GP), needles.
3. Benchtop microcentrifuge (Eppendorf™).

2.8 SAXS Data Analysis

1. Software required: ATSAS suite (*see* **Note 4**).

2.9 SAXS-Aided RNA Structure Prediction and Computational Modeling

1. Software required: ModeRNA (*see* **Note 5**), Rosetta (*see* **Note 6**), Xplor-NIH (*see* **Note 7**).

2.10 RNA Ensemble Analysis Using the Ensemble Optimization Method

1. Software required: Xplor-NIH, EOM (*see* **Note 8**).

3 Methods

3.1 General Considerations

Structural modularity is one of the general features of lncRNA, which often consists of several independently folded subdomains connected with flexible linkers [7]. The subdomain boundaries can be derived based on its secondary structural domains map [21]. The secondary structure of ZIKV sfRNA was recently determined by SHAPE chemical probing, which reveals four subdomains (xrRNA1, xrRNA2, ψDB12, and 3′SL) [7]. Due to its large size (404 nts), a divide-and-conquer strategy was employed for structural characterization of the ZIKV sfRNA in solution by SAXS (*see* Fig. 1a). Optimal Mg^{2+} concentration for proper folding of the complete ZIKV sfRNA was firstly determined, since Mg^{2+} is generally essential to RNA folding and structure. The subconstructs for individual and combined subdomains of ZIKV sfRNA were constructed, in vitro transcribed, purified under nondenaturing conditions (*see* **Note 9**), and studied at the optimal Mg^{2+} concentration by SAXS. SAXS data analysis yields overall structural parameters, folding properties and low-resolution shape structures for the individual and combined subdomains. 3D structural models for the individual or combined subdomains were built up from the secondary structure with the aid of SAXS data (*see* Fig. 2). With the availability of 3D atomic models for the respective individual or combined subdomains, a 3D atomic model was generated with a SAXS-restrained rigid-body modeling algorithm and the accessible 3D conformational ensemble was characterized for the complete ZIKV sfRNA (*see* Fig. 2).

3.2 DNA Molecular Cloning

1. Order the cDNA encoding the full length lncRNA (here the ZIKV sfRNA) with an immediate upstream T7 promoter from a gene synthesis company. The cDNA is usually inserted into a plasmid vector.
2. Design and order DNA oligonucleotides (as both forward and reverse primers) for PCR cloning.
3. Molecular cloning of the lncRNA subconstructs using the above plasmid vector as template [22] (*see* Fig. 1b).

3.3 RNA Production by In Vitro Transcription

1. Design and order oligonucleotides for template DNA amplification (*see* **Note 2,** Fig. 1b).
2. Amplify the template DNA fragment encoding each specific subconstruct and full length of lncRNA by standard PCR reaction.
3. Set up small-scale transcription reactions (20 μl) for each particular subconstruct and the full length of lncRNA (*see* Table 1).
4. Incubate for 1–2 h at 37 °C.

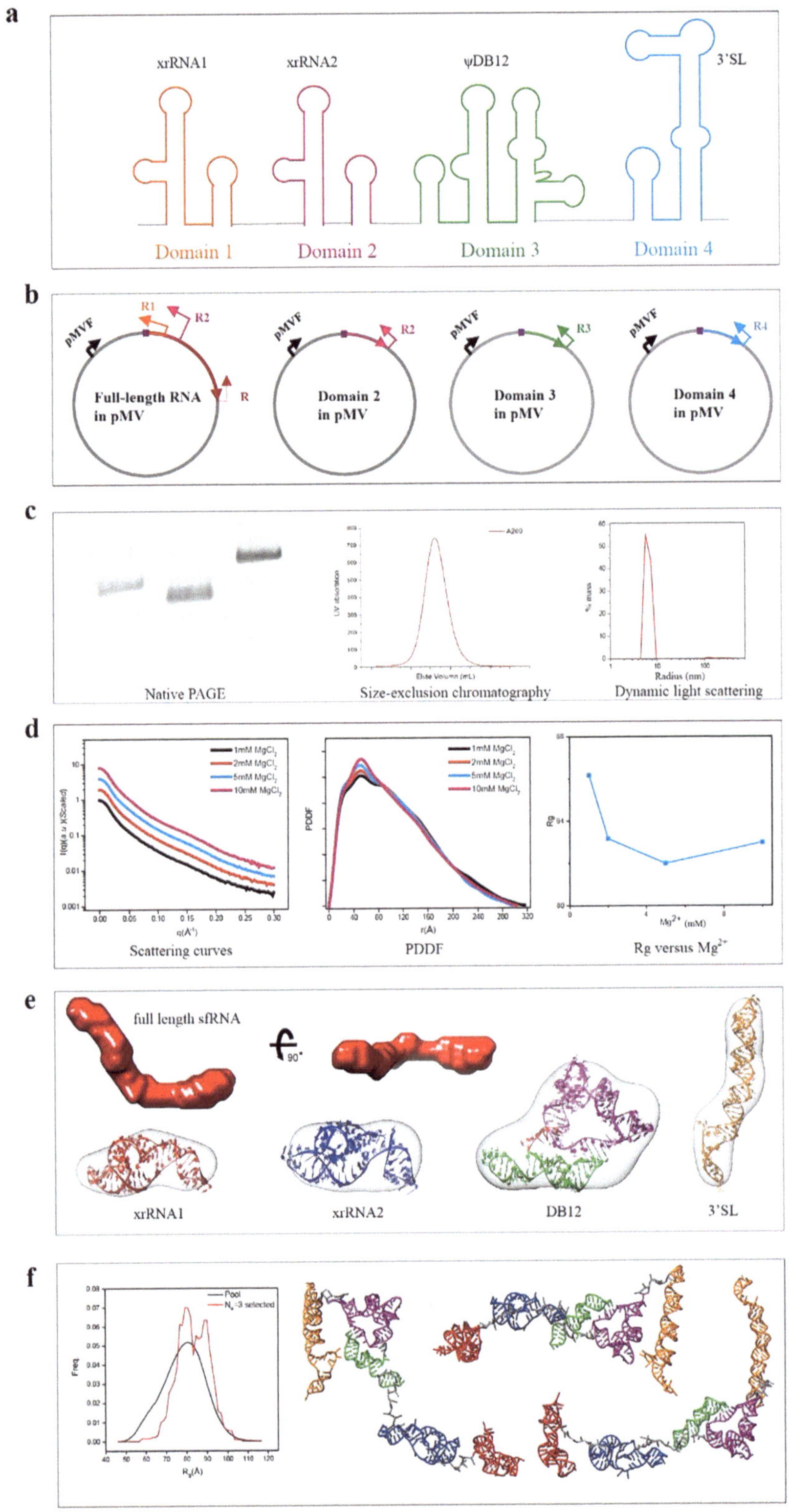

Fig.1 3D structural analysis of a lncRNA, the sfRNA from ZIKV, by combined use of SAXS and computational modeling. (**a–f**) Starting from the primary sequence and secondary structures of the full length ZIKV sfRNA (**a**),

5. Run the transcription mixture on a native polyacrylamide gel. The type of gel depends on the size of the transcript to be purified, but in most cases, a 5–10% polyacrylamide gel will be appropriate.
6. Vary the concentrations of Mg^{2+} and/or DNA template to optimize the transcription condition for each particular RNA (*see* Table 1).
7. Carry out large-scale transcription reactions (2–20 ml) for the RNA to be purified under the optimized transcription condition. Stop the reaction after 3 h by adding 10 mM EDTA.
8. Aliquot 10 μl of the large-scale reaction mixture and run on a native polyacrylamide gel. Note the migration of the target RNA relative to the tracking dyes bromophenol blue and xylene cyanol on the gel.

3.4 RNA Purification by Preparative Chromatography

1. Select the chromatography column according to the band separation and the size of the RNA to be purified on the native polyacrylamide gel.
2. Prepare the column according to the manufacturer's recommendation.
3. Centrifuge the transcription reaction mixture at 12,000 g to remove the precipitate. When needed, concentrate the reaction mixture using the Amicon Ultra Centrifugal Filters (*see* **Note 10**) or ethanol precipitation [23].
4. Load the reaction mixture to the HPLC system using appropriate sample loops and run the system.
5. Collect the elute containing the target transcript with each fraction of 0.5 ml volume.
6. Run the fractions on a native polyacrylamide gel. Collect, combine, and concentrate the fractions of target RNA that are homogenous.

Fig.1 (continued) a "divide-and-conquer" strategy was employed to construct the plasmid vectors for the individual domains of ZIKV sfRNA (**b**). RNAs were prepared by in vitro transcription and purified under native conditions. The purity and homogeneity of RNA samples were checked and conditioned for SAXS analysis (**c**). The optimal Mg^{2+} concentration for full length sfRNA folding was determined by analyzing SAXS data collected at various Mg^{2+} concentrations, and SAXS data for the individual subdomains were collected at the optimal Mg^{2+} concentration (**d**). Ab initio 3D shape structures were reconstructed for the full length sfRNA and its individual subdomains. The 3D atomic models were predicted with the aid of SAXS data and fitted into the ab initio reconstructed shape envelopes from SAXS data (**e**). Atomic model for full length sfRNA was assembled from the atomic models of the individual subdomains by SAXS-restrained rigid-body modeling using Xplor-NIH. A large conformational pool was generated using Xplor-NIH and ensemble refinement and flexibility analysis was performed using Ensemble Optimization Method (EOM) (**f**)

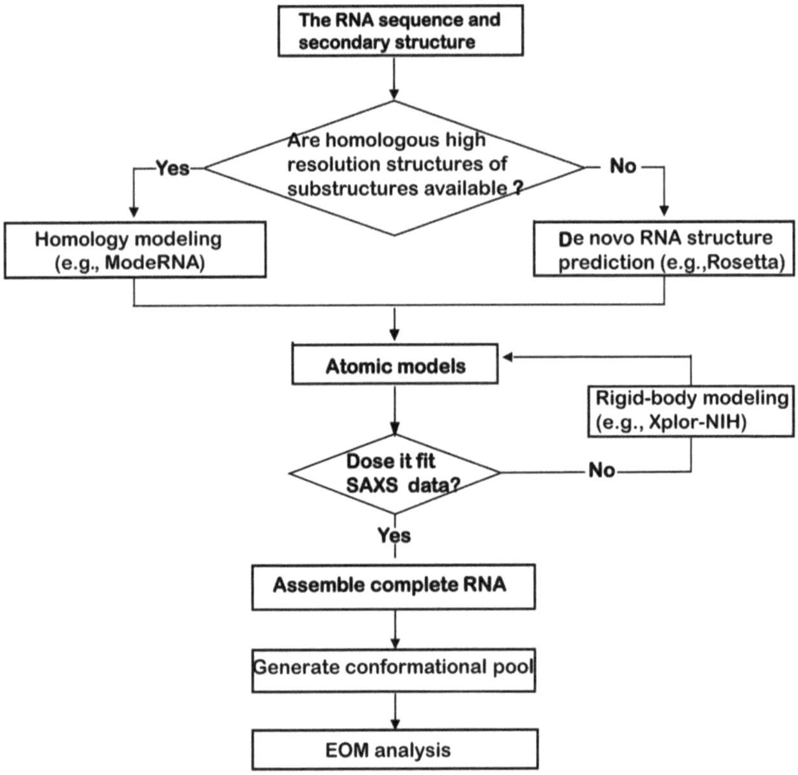

Fig. 2 Flowchart for SAXS-aided computational modeling of large RNAs starting from the primary sequence and secondary structure information

Table 1
Components of the RNA in vitro transcription reaction

Components	Final concentration[a]	Volume required (μL) 20 μL reaction[b]	Volume required (μL) 5 mL reaction
1 M Tris-HCl, pH 8.0	40 mM	0.8	200
0.2 M $MgCl_2$	5 mM–40 mM	0.5–4	125–1000
0.1 M Spermidine	2 mM	0.4	100
1 M Dithiothreitol	10 mM	0.2	50
40 mM each rNTPs	4 mM	2	500
PCR product	–	2–8	500–2000
T7 RNA polymerase	0.1 mg/mL	1	250
H_2O	–	Up to 20	Up to 5000

[a]Such condition produces good yield for most RNAs but optimization of PCR product and/or $MgCl_2$ concentrations may be needed for each particular RNA construct

[b]For a small-scale transcription reaction (20 μL), it is better to start from the 10 × transcription buffer containing 400 mM Tris-HCl pH 8.0, 100 mM DTT, 20 mM spermidine, avoiding mixing the components individually

3.5 RNA Purification by Preparative Native PAGE

1. Clean glass plates, spacers, and combs. Assemble gel plates and seal the sides carefully using high-vacuum grease and binder clips.
2. Prepare 130 ml of gel and pour into the assembled gel plate slowly, remove air bubbles, insert combs, and let it polymerize horizontally overnight.
3. Remove the bottom spacer and comb, and mount the gel plate to the electrophoresis chamber. Rinse the well thoroughly with 0.5 × TBM.
4. Mix the transcription reaction solution with loading dye, and load the mixture to the well.
5. Perform the electrophoresis at 350 V/50 mA at 4 °C. To ensure an optimal separation, run the electrophoresis as long as possible. Use the tracking dyes to infer the migration of the target RNA and stop the electrophoresis before it migrates outside the gel.
6. Remove the gel from the plates and cover it with a piece of clean plastic wrap.
7. Place the gel on the TLC plate and illuminate with the UV light source (254 nm). The target RNA will appear as a dark band. Cut the gel band using a clean scalpel blade.
8. Extract the target RNA at 4 °C using the Crush-and-Soak method. Concentrate the RNA using the Amicon Ultra Centrifugal Filters with appropriate MW cutoff.

3.6 RNA Quantification with UV-Vis Spectroscopy

1. Measure the OD260 of each particular RNA using a Nanodrop 2000 or any other UV-vis spectrometer. Dilute or concentrate the RNA solution to reach OD260 in the range of 0.2–0.5 if needed (*see* **Note 11**).
2. Calculate the molar extinction coefficient at 260 nm for each particular RNA from its primary sequence using the web server of Oligo Calc (*see* **Note 12**).
3. Calculate the molar concentration (μM) and/or mass concentration (mg/ml) of the RNA solution.

3.7 Sample and Buffer Preparation for SAXS

3.7.1 General Considerations

The solution scattering profile of a macromolecule from a SAXS experiment is plotted as the scattering intensity against momentum transfer, q, where $q = 4\pi \sin \theta/\lambda$, 2θ is the scattering angle, and λ is the wavelength of X-ray used. In practice, the scattering profile of the solvent used to dissolve the macromolecule is recorded separately and subtracted from that of the solution of macromolecule, resulting in the scattering profile of macromolecules in solution. As the SAXS curves usually rapidly decay with angle and the scattering intensity at higher angles is usually much weaker, careful sample and buffer preparation and obtaining an exact matching between buffer and sample solution are of critical importance for accurate background subtraction in data processing and high-quality SAXS data.

SAXS data can be collected with laboratory or synchrotron-based X-ray sources. In this chapter, we limit the discussion for planning SAXS measurements at beamlines using a synchrotron source. The samples are generally prepared in the local laboratory and transported to a remote synchrotron beamline. While the specific requirements may vary across different beamlines, there are common factors to be considered, such as sample purity, polydispersity, quantity, stability, and sample shipment.

3.7.2 RNA Sample Preparation

1. The purity and homogeneity of RNAs are crucial for SAXS data interpretation (*see* **Note 13**). It is important to check the integrity and homogeneity of RNA samples prior to SAXS experiments. A number of techniques are especially suited for detecting and quantifying higher oligomeric species, including native polyacrylamide gel electrophoresis, analytical size-exclusion chromatography, dynamic light scattering, and analytical ultracentrifugation (*see* Fig. 1c). Generally, a single tight band or symmetric peak indicates high purity and monodispersity of the RNA sample.
2. Mg^{2+} is known to facilitate and stabilize the folding of RNAs [24]. It is recommended to determine the optimal Mg^{2+} concentration for lncRNA folding. Usually, lncRNA at various Mg^{2+} concentrations (0.01, 0.1, 0.2, 0.5, 1, 2, 5, 10, 20 mM) are prepared (*see* Fig. 1d).
3. Samples for lncRNA subconstructs at the optimal Mg^{2+} concentration are prepared.
4. A dilution series (stock, twofold, and fourfold dilution) of measurements for lncRNA and its subconstructs at each condition are recommended. The dilution can be prepared in advance or onsite.
5. Each SAXS measurement will require ~100 μL of sample at concentrations between 0.3 and 1.0 mg/ml.

3.7.3 Buffer Recommendation

1. X-ray-induced radiation damage is a major obstacle of SAXS as it can cause aggregation, fragmentation, or unfolding of macromolecules. To minimize radiation damage, small-molecule free radical scavengers or polyols can be included in the buffer. An effective buffer can contain 20–50 mM of organic buffer agents (such as Tris-HCl, HEPES), 1–5 mM of reducing agents (ascorbic acid, dithiotheritol, TCEP, β-mercaptoethanol), 100–500 mM of salts (NaCl, KCl), and 3–5% Glycerol.

2. Avoid high concentrations of small molecules (e.g., 2 M KCl) or the addition of electron-dense molecules (30–50% sucrose) in the buffer, which can reduce the contrast of the solution.
3. Provide sufficient buffer (for example 50 ml) as the beamline suggests.

3.7.4 Sample and Buffer Matching

Several approaches can be taken to prepare the exact matching buffer, which is crucial for any SAXS measurement.

1. The best way to obtain the exact matching buffer is to perform dialysis. However, it may take days to complete the process of buffer exchange by dialysis.
2. Buffer exchange can also be performed with a desalting column or by size-exclusion chromatography. The running buffer should be kept as the reference buffer in SAXS measurement.
3. Ultrafiltration is an alternative option for obtaining a matched buffer quickly. Instead of relying on passive diffusion in dialysis, the solutions are forced through the membrane by pressure or centrifugation in ultrafiltration, thus the process of buffer exchange can be accelerated. Ultrafiltration devices with different MWCO are commercially available.

3.7.5 Sample Shipments

1. Once the samples and buffers are obtained with high quality and sufficient amount, final shipment arrangements can be made. Discuss the receiving address, and instructions on the storage conditions with the local beamline staff.
2. The lncRNA samples should be shipped to the beamline utilizing dry ice or cold packs, depending on RNA stability. All the tube caps should be carefully sealed with parafilm during shipment.
3. Do not forget to ship the matching buffers with the samples, which can be shipped at room temperature.
4. Consumable supplies (dialysis accessories, syringes and needles, ultrafiltration devices, etc.) should also be shipped with the samples.

3.8 SAXS Data Collection

1. Consult the beamline staff and follow the instructions on sample handling at the beamline. While the specific instruction may be different, there are common points to be considered.
2. Get familiar with the basic experimental setup, the instrument control, and data analysis software at the beamline.
3. Centrifuge the samples at high speed (12,000 g) for at least 10 min or pass the sample solution through a 0.22 μm syringe filter prior to the experiment. If precipitates can be seen at the bottom of the tube, transfer the sample solution to a new tube.

4. It is recommended to measure the samples from low to high concentrations. Run buffer measurements before and after each sample measurement. To remove any residuals from previous runs, wash the flow cell extensively after each sample measurement.
5. Collect 20–30 individual 1-s exposures for each buffer or sample measurement. Reduce the 2D image to 1D scattering curve using the beamline-specific control script. Examine the superimposition of the 1D scattering curves to check if radiation damage occurs immediately after each sample measurement. If so, repeat data collection at a reduced exposure time or lower concentration, which can sometimes minimize radiation damage.
6. Remove the outliers from the superimposed scattering curves, average the remaining scattering curves for the RNA samples and buffers, respectively, using the beamline-specific script, and save the averaged scattering curves for each buffer or RNA sample.

3.9 SAXS Data Analysis

Among the many software available for SAXS data analysis, the ATSAS package is the most popular. Below we briefly outline the procedures of data analysis using the ATSAS suite of programs.

1. Perform background subtraction of the averaged buffer scattering curve from the averaged RNA sample scattering curve using Primus. Save the scattering profile for the RNA sample at each RNA concentration.
2. Extrapolate to zero concentration to obtain a scattering profile that represents infinite dilution using Primus.
3. Perform Guinier plot for the scattering profile of the RNA sample at infinite dilution. Make Guinier fitting and calculate the R_g, I_0 in the range of $qR_g < 1.3$ for the RNA sample at each Mg^{2+} concentration. Evaluate the sample quality based on the features in Guinier plots.
4. Make dimensionless Kratky plot, $(qR_g)^2 I(q)/I_0$ vs. qR_g, with the RNA scattering curve at infinite dilution to evaluate RNA folding and flexibility (*see* **Note 14**).
5. Calculate PDDF with the scattering data in the q range no larger than 0.30 $Å^{-1}$ using GNOM. Note the R_g, I_0 and D_{max} (*see* Fig. 1d).
6. Plot the R_g and D_{max} against Mg^{2+} concentration to determine the optimal Mg^{2+} concentration for RNA folding (*see* Fig. 1d).
7. Calculate the molecular weight of the RNA using the power-law parameter Q_R (V_c^2/R_g), which is derived from the SAXS invariant of volume-of-correlation (V_c) to check the RNA oligomeric state [19].

8. Run ab initio shape reconstruction for the RNA using the program DAMMIN (*see* Fig. 1e). Generally, 20–40 bead models are initially generated using the scattering data in the q range of 0.006–0.30 $Å^{-1}$. Align, average, and filter the 20–40 reconstructions using the program DAMEVER from the ATSAS suite to produce a final low-resolution bead model. The DAMEVER log file will list the normalized spatial discrepancy (NSD) for each model after comparison with all other models. NSD reflects the quantitative similarity between sets of three-dimensional models. NSD < 0.7 indicates the high similarity of each model to those against which it was compared.

3.10 RNA 3D Structure Prediction and SAXS-Aided Computational Modeling

1. Prepare the primary sequences of lncRNA and its subdomains in a text file. Prepare the inputs of RNA secondary structure using the dot bracket notation, based on the secondary structure map of lncRNA determined by SHAPE chemical probing (*see* Fig. 2).
2. If homologous high-resolution structures are available for the subdomains of lncRNA, build up a homology model with the program ModeRNA using the high-resolution structure as template.
3. If no homologous models are available for the subdomains of lncRNA, generate ensembles of de novo all-atom 3D models for the subdomains first using the program Rosetta [19]. Models in the ensembles are further screened against the experimental SAXS data, and the ones which have best fitting to the experimental SAXS data and can be nicely superimposed onto the bead models are chosen as the best representative all-atom 3D model for the respective constructs.
4. With the availability of atomic models for the individual subdomains, rigid-body modeling can be carried out to construct atomic models for the combined subdomains or the complete lncRNA using the Xplor-NIH package [25], during which the respective individual subdomains are kept as rigid bodies, the linkers between are allowed to translate or rotate freely, and a simulated annealing algorithm is performed to optimize the best position and orientation of the individual domains against SAXS data.
5. Fit the atomic models to the respective ab initio 3D shapes of individual subdomains (see Fig. 1e).

3.11 Ensemble Analysis Using the Ensemble Optimization Method

1. Generate a conformational pool containing a large set of random models for the lncRNA, ZIKV sfRNA using the Xplor-NIH package (*see* Fig. 2), which follows a simulated annealing protocol driven by molecular dynamics simulation in torsion angle space that is subject to a target function comprised of

bond length, bond angles, improper dihedral angles, which specify chirality and planarity of functional groups, a quartic van der Waals repulsion and R_g terms to prevent atomic overlap, a multidimensional torsion-angle database potential to improve backbone and sidechain conformation. During the simulation, subdomains of xrRNA1, xrRNA2, ψDB12, and 3′SL of ZIKV sfRNA are treated as independent rigid bodies, and the linkers between are allowed to translate and/or rotate freely. To minimize the effect of the starting model conformation on sampling, two round calculations are carried out. 120 models are generated by MD simulation in the first round, among which 20 models with high total energy and bond angle energy terms are picked out. The remaining 100 models are used as starting models for the next round calculation, which results in a total of 12,000 models. The same procedure as the first round is used to pick out those models with high total energy and bond angle energy terms, and a total of 10,000 models is left to assemble a conformational pool for ZIKV sfRNA.

2. A minimal sub-ensemble of the above conformational pool that collectively reproduces the SAXS profile of ZIKV sfRNA using the genetic algorithm GAJOE is optimized using the EOM program. The size of the minimal ensemble was varied from 1 to 20 (1, 2, 3, 4, 6, 8, 20) to test the effect of ensemble size on the quality of the fitting, and the R_g distribution of the selected ensemble can be compared with that of the random pool which gives an estimation of flexibility (*see* Fig. 1f).

4 Notes

1. To study the individual and combined subdomains of ZIKV sfRNA, another three subclones encoding xrRNA2, ψDB12, and 3′SL, respectively, are made from the plasmid vector encoding full length lncRNA, which can be used to prepare DNA template for in vitro transcription of complete sfRNA, individual domain of xrRNA1, and combined domains of xrRNA1 and xrRNA2.
2. The forward oligonucleotide primer is designed to target the plasmid DNA sequence 500 bp upstream of the T7 promoter, the reverse oligonucleotide primers are designed to target the 3′ ends of each particular RNA transcript. Such design results in template DNA significantly larger than the RNA transcript, enabling good separation of the template DNA by native PAGE gel or size exclusion chromatography [26].
3. The T7 RNA polymerase can be produced in-house using established protocols [27]. To remove any potential RNase

contamination from the *E.coli* debris, a further SEC purification step is recommended after the standard nickel-affinity purification technique.

4. ATSAS suite is a widely used package in the bioSAXS community and freely available to academics at https://www.embl-hamburg.de/biosaxs/software.html. The website also provides detailed manuals and course lectures. Other packages or programs may be used as well, such as SCATTER, BioXTAS RAW, and Igor Pro (http://gisaxs.com/index.php/Software).
5. ModeRNA is a program for comparative modeling of RNA 3D structures. It requires a pairwise sequence alignment and a structural template to generate a 3D structural model of the target RNA sequence [28].
6. Rosetta uses the Fragment Assembly of RNA (FARNA) algorithm to model RNA structure, when the secondary structure is specified. For larger RNAs, fragment assembly without refinement appears more efficient. Although atomic accuracy is unlikely, models accurate at nucleotide or helix resolution can be achieved, especially with constraints from experiments [19].
7. Xplor-NIH is a flexible biomacromolecular structure determination program [29]. Experimental restraints from solution and solid state NMR, electron paramagnetic resonance (EPR), and X-ray scattering experiments can be accommodated during structure calculations. Xplor-NIH can also be used for ensemble calculations where the experimental data cannot be accounted for by a unique structure.
8. EOM is a part of ATSAS suite of programs, which has been widely used to describe flexible proteins. The EOM approach consists of three steps: (1) Generation of a large pool of conformers representing the conformational space sampled by the molecules in solution, (2) Selection of a minimal ensemble of conformers that collectively describes the SAXS data, (3) Description of the structural properties of the selected ensemble in terms of size distribution and flexibility. Xplor-NIH can be used to generate the conformational pool for lncRNA, then the program GAJOE in EOM is run for selection and description of the sub-ensemble of RNA.
9. The refolding process of large RNA is complicated. It is suggested to purify large RNA including lncRNA under native conditions [30].
10. The membrane permeability of a molecule depends upon its shape, degree of hydration, ionic charge, and polarity. It is recommended to select a MWCO that is half the size of the MW of the species to be retained.

11. The absorption of RNA at 260 nm is proportional to concentration according to Beer Lambert's law. However, this is not followed in higher or lower concentrations of the RNA.
12. Oligo Calc (http://biotools.nubic.northwestern.edu/OligoCalc.html) is a free, convenient web interface for calculating the physical properties of DNA and RNA including molecular weight and molar extinction coefficients.
13. Usually, RNA with 95% purity can be achieved by native chromatography or gel electrophoresis purification. However, higher oligomeric species resulting from misfolded or nonspecifically aggregated RNA or smaller degraded species may form during concentration or storage, which are detrimental to SAXS data quality.
14. The dimensionless Kratky plot is a very useful method to infer the folding property of a molecule in solution from its shape. The Kratky plot of a globular and compact molecule has a typical bell shape with a clear maximum, but no such maximum can be observed for a completely unfolded molecule. Instead, the curve displays a plateau or increase at elevated qR_g.

Acknowledgments

This work was supported by grants from the National Natural Science Foundation of China (No. U1832215), the Beijing Advanced Innovation Center for Structural Biology, and Beijing Frontier Research Center for Biological Structure.

References

1. Ma L, Bajic VB, Zhang Z (2013) On the classification of long non-coding RNAs. RNA Biol 10:925–933
2. Statello L, Guo CJ, Chen LL et al (2020) Gene regulation by long non-coding RNAs and its biological functions. Nat Rev Mol Cell Biol 22(2):96–118
3. Arun G, Diermeier SD, Spector DL (2018) Therapeutic targeting of long non-coding RNAs in cancer. Trends Mol Med 24:257–277
4. Blythe AJ, Fox AH, Bond CS (2016) The ins and outs of lncRNA structure: How, why and what comes next? Biochim Biophys Acta 1859: 46–58
5. Zampetaki A, Albrecht A, Steinhofel K (2018) Long non-coding RNA structure and function: is there a link? Front Physiol 9:1201
6. Mcfadden EJ, Hargrove AE (2016) Biochemical methods to investigate lncRNA and the influence of lncRNA:protein complexes on chromatin. Biochemistry 55:1615–1630
7. Zhang Y, Zhang Y, Liu ZY et al (2019) Long non-coding subgenomic flavivirus RNAs have extended 3D structures and are flexible in solution. EMBO Rep 20:e47016
8. Brosey CA, Tainer JA (2019) Evolving SAXS versatility: solution X-ray scattering for macromolecular architecture, functional landscapes, and integrative structural biology. Curr Opin Struct Biol 58:197–213
9. Fang X, Stagno JR, Bhandari YR et al (2015) Small-angle X-ray scattering: a bridge between RNA secondary structures and three-dimensional topological structures. Curr Opin Struct Biol 30:147–160
10. Chen Y, Pollack L (2016) SAXS studies of RNA: structures, dynamics, and interactions with partners. Wiley Interdiscip Rev RNA 7: 512–526

11. Fang X, Wang J, O'carroll IP et al (2013) An unusual topological structure of the HIV-1 Rev response element. Cell 155:594–605
12. Fang X, Michnicka M, Zhang Y et al (2017) Capture and release of tRNA by the T-loop receptor in the function of the T-Box riboswitch. Biochemistry 56:3549–3558
13. Zuo X, Wang J, Foster TR et al (2008) Global molecular structure and interfaces: refining an RNA:RNA complex structure using solution X-ray scattering data. J Am Chem Soc 130:3292–3293
14. Wang J, Zuo X, Yu P et al (2009) A method for helical RNA global structure determination in solution using small-angle x-ray scattering and NMR measurements. J Mol Biol 393:717–734
15. Somarowthu S (2016) Progress and current challenges in modeling large RNAs. J Mol Biol 428:736–747
16. Ponce-Salvatierra A, Astha, Merdas K et al (2019) Computational modeling of RNA 3D structure based on experimental data. Biosci Rep 39:BSR20180430
17. Charley PA, Wilusz J (2016) Standing your ground to exoribonucleases: function of flavivirus long non-coding RNAs. Virus Res 212: 70–77
18. Chaudhuri B, Muñoz IG, Qjan S, Urban V (2017) Biological small angle scattering: techniques, strategies and tips. Springer, Singapore
19. Cheng CY, Chou FC, Das R (2015) Modeling complex RNA tertiary folds with Rosetta. Methods Enzymol 553:35–64
20. Magnus M, Matelska D, Lach G et al (2014) Computational modeling of RNA 3D structures, with the aid of experimental restraints. RNA Biol 11:522–536
21. Somarowthu S, Legiewicz M, Chillon I et al (2015) HOTAIR forms an intricate and modular secondary structure. Mol Cell 58:353–361
22. Hoseini SS, Sauer MG (2015) Molecular cloning using polymerase chain reaction, an educational guide for cellular engineering. J Biol Eng 9:2
23. Green MR, Sambrook J (2020) Precipitation of RNA with ethanol. Cold Spring Harb Protoc 2020:101717
24. Niu X, Liu Q, Xu Z et al (2020) Molecular mechanisms underlying the extreme mechanical anisotropy of the flaviviral exoribonuclease-resistant RNAs (xrRNAs). Nat Commun 11: 5496
25. Schwieters CD, Kuszewski JJ, Tjandra N et al (2003) The Xplor-NIH NMR molecular structure determination package. J Magn Reson 160:65–73
26. Wang Y, Kathiresan V, Chen Y et al (2020) Posttranscriptional site-directed spin labeling of large RNAs with an unnatural base pair system under non-denaturing conditions. Chem Sci 11:9655–9664
27. Rio, DC (2013) Expression and purification of active recombinant T7 RNA polymerase from E. coli. Cold Spring Harbor Protocols 11: pdb. prot078527
28. Rother M, Rother K, Puton T et al (2011) ModeRNA: a tool for comparative modeling of RNA 3D structure. Nucleic Acids Res 39: 4007–4022
29. Schwieters CD, Bermejo GA, Clore GM (2018) Xplor-NIH for molecular structure determination from NMR and other data sources. Protein Sci 27:26–40
30. Chillon I, Marcia M, Legiewicz M et al (2015) Native purification and analysis of long RNAs. Methods Enzymol 558:3–37

Chapter 11

Combining Biophysical Methods for Structure–Function Analyses of RNA in Solution

Yun-Tzai Lee, Lixin Fan, Jienyu Ding, and Yun-Xing Wang

Abstract

RNA-level regulation by riboswitches relies on the specific binding of small metabolites to the aptamer domain to trigger substantial conformational changes that affect transcription or translation. Although several biophysical methods have been employed to study such RNAs, the utility of any one single method is limited. Hybrid approaches, therefore, are essential to better characterize these intrinsically dynamic molecules and elucidate their regulatory mechanisms driven by ligand-induced conformational changes. This chapter outlines procedures for biochemical and biophysical characterization of RNA that employs a combination of solution-based methods: isothermal titration calorimetry (ITC), small-angle X-ray scattering (SAXS), and atomic force microscopy (AFM). Collectively, these tools provide a semi-quantitative assessment of the thermodynamics associated with ligand binding and subsequent conformational changes.

Key words RNA structure and dynamics, Riboswitch, Atomic force microscopy, Isothermal titration calorimetry, Small-angle x-ray scattering

1 Introduction

Although ITC and SAXS provide bulk samples rather than single-molecule measurements, they have the advantages of being label-free methods and recorded under solution conditions. ITC provides direct measurements of binding enthalpy, stoichiometry, RNA entropy changes, and thermodynamic populations. SAXS, on the other hand, allows for the study of the structure and dynamics of RNA under varying buffer conditions, such as salt concentration, pH, with or without ligand, as well as under changing sample environments, such as temperature and pressure [1–3]. SAXS data provide not only global information about the size and shape of the RNA but also the extent of its structural variation manifested as an ensemble of conformers, and can provide information on individual conformer populations (volume fractions) [4–7]. SAXS data can also be used in tandem with other biophysical methods (including

Jienyu Ding et al. (eds.), *RNA Structure and Dynamics*, Methods in Molecular Biology, vol. 2568,
https://doi.org/10.1007/978-1-0716-2687-0_11,

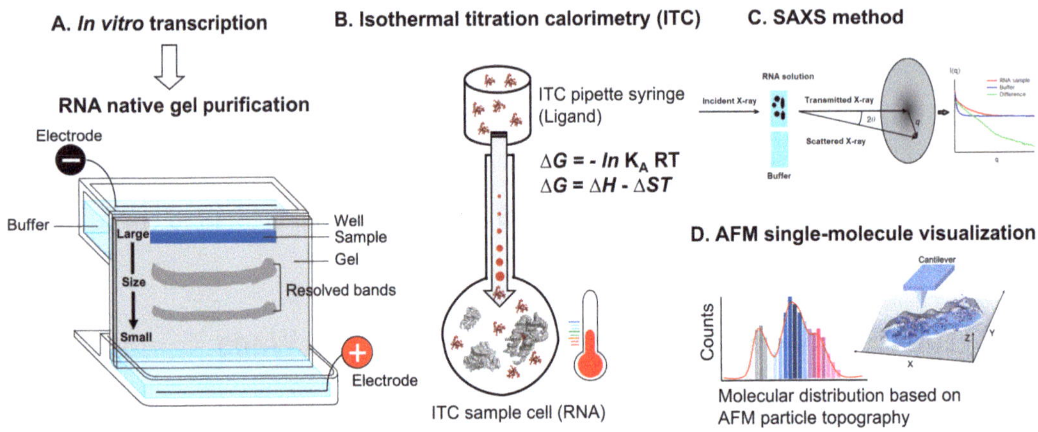

Fig. 1 Workflow for using hybrid assessments to characterize ligand binding to RNA and its associated conformational changes. (**a**) RNA purification using gel electrophoresis. (**b**) Pictorial representation of ITC instrument and derived thermodynamic terms. (**c**) Illustration of how SAXS generates a scattering profile of molecules in bulk-solution. (**d**) Statistical distribution of conformers in the AFM topography, classified by physical properties (*e.g.*, size, shape, volume)

crystallography, NMR, AFM, and cryo-EM) by providing additional restraints that further improve simulations and validate structural models [8–13]. The third technique in the combined approach is AFM, which has the advantage of direct visualization and characterization of RNA at the single molecule level. Moreover, AFM data can be recorded using nanomolar RNA concentrations under physiologically relevant solution conditions. Here, we present this hybrid method combining ITC, SAXS, and AFM single molecule visualization (Fig. 1) to study the conformational and heterogeneity of the *Thermoanaerobacter tengcongensis* (*Tte*) cobalamin (Cbl) riboswitch aptamer domain (*Tte*AdoCbl) [14].

AFM single-particle visualization of *Tte*AdoCbl revealed a mixture of monomeric and dimeric species, each consisting of various conformers (Fig. 2). A few higher order oligomers were also observed. Thus, it is important to validate the structures using different techniques. Both SAXS and AFM can be used to characterize RNA in solution, and SAXS allows studying ensemble behaviors of heterogeneous conformations under different buffer conditions [15–17]. Therefore, SAXS is a suitable tool to validate structures derived from AFM. We developed a method that synthesizes the contributions of all conformational structures derived by AFM and compares synthesized results with experimental SAXS data. Singular value decomposition (SVD) analysis has been used to separate elementary contributions of optical spectra [18] and X-ray diffraction patterns [19] in various biophysical applications. We implemented this method to analyze ITC thermograms acquired for the titration of *Tte*AdoCbl with ligand. By conducting SVD analysis, the raw thermogram can be deconvoluted into

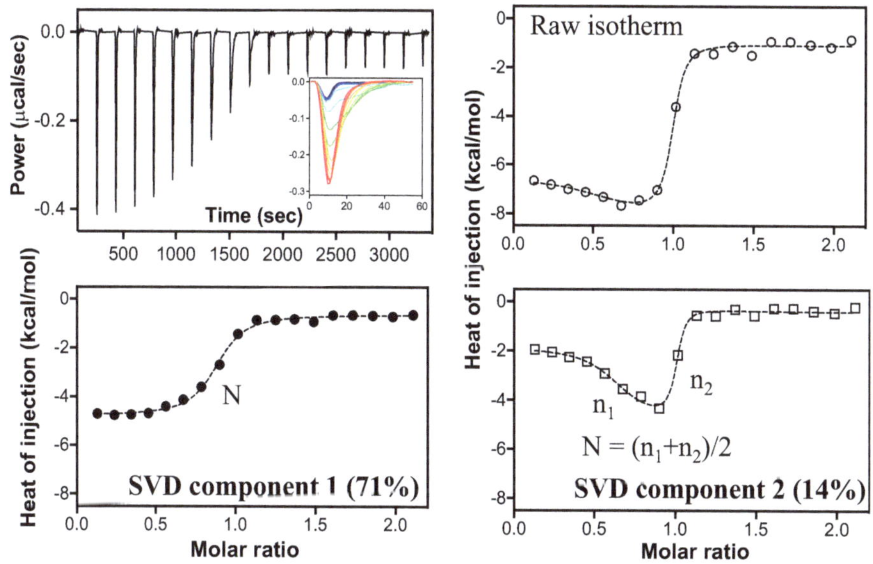

Fig. 2 ITC thermograms and integrated isotherms for *Tte*AdoCbl (50 μM) titrated with AdoCbl (500 μM). (**a**) Raw data after baseline subtraction, which exhibits unconventional peak broadening (inset panel). (**b**) Fitted raw isotherm, which is not a monotonic sigmoidal curve, suggesting that AdoCbl binding triggers a substantial conformational change through an induced-fit binding scenario, which is consistent with multiple conformations observed by AFM and SAXS ensemble fitting (*see* Fig. 3). (**c–d**) Deconvoluted isotherms for principal components 1 (71%) and 2 (14%) after SVD analysis. N is the binding stoichiometry derived from the fitting of the isotherm associated with SVD component 1. n1 and n2 are the two binding stoichiometries determined using the two-site binding model, associated with SVD component 2. The average of the two values is as much as N as a global sharing factor

principal components that are statistically significant and whose thermodynamic parameters may be related to different populations observed by AFM and SAXS ensemble analysis. Together, these methods provide a semi-quantitative assessment of RNA conformations under various conditions, including the absence or presence ligand, and a means to validate and interpret data in ways that are not possible by any singular approach.

2 Materials

All buffers should be prepared using the highest quality-grade reagents (RNase-free where applicable) using ultrapure deionized water that has been treated with 0.1% v/v diethylpyrocarbonate (DEPC) and autoclaved twice for 40 min each (DEPC-water). Filter all buffers through a 0.22-μm sterile filter. All containers and equipment that come into direct contact with RNA should be first rinsed with DEPC-water to avoid RNase contamination (*see* **Note 1**).

2.1 Instrumentation

1. Isothermal microcalorimeter: Microcal iTC200 or VP-ITC (Malvern Panalytical Ltd, United Kingdom), or equivalent.
2. High-speed atomic force microscope, including relevant hardware and consumables (*e.g.*, cantilevers, tubing). Here we use a Cypher VRS (Asylum Research, Oxford Instruments).
3. UV-VIS spectrometer.
4. Synchrotron or in-house X-ray source for SAXS data acquisition, preferably with a PILATUS (Dectris AG, Baden-Daettwil, Switzerland) detector.
5. Refrigerated centrifuge.
6. High-voltage electrophoresis power supply.
7. Vertical slab gel electrophoresis system (Fig. 1a).
8. UV-transilluminator lamp with dual wavelength (254 and 365 nm).

2.2 In Vitro Transcription (IVT)

1. IVT reaction buffer stock solutions: 1 M HEPES buffer pH 7.5, 100 mM spermidine, 1 M $MgCl_2$, dithiothreitol (DTT) powder (*see* **Note 2**).
2. 100 mM ribonucleoside triphosphate (rNTP) premixed solution.
3. Synthesized double-stranded DNA template with T7 promoter sequence at the 5′-terminus (*see* **Note 3**).
4. Recombinant T7 RNA polymerase (*see* **Note 4**).
5. Yeast inorganic pyrophosphatase (optional).
6. RNA elution buffer: 300 mM sodium acetate (NaOAc), pH 5.3, 0.1 mM ethylenediaminetetraacetic acid (EDTA).
7. RNase inhibitor cocktail (*e.g.*, SUPERaseInTM, Ambion, Inc. Austin, TX).

2.3 RNA Purification Using Native Polyacrylamide Gel Electrophoresis

1. 30% Bis-acrylamide solution.
2. 10× Tris-borate EDTA (TBE) buffer pH 7.5.
3. 10% w/v ammonium persulfate solution.
4. TEtraMethylEthyleneDiamine (TEMED).
5. 2× RNA native gel loading dye: 10 mM Tris HCl pH 7.5, 0.025% (w/v) bromophenol blue, 0.5 mM EDTA, 20% glycerol, (store at −20 °C).
6. Centrifugal filter unit (*e.g.*, Amicon® Ultra, MilliporeSigma, Burlington, Massachusetts) with nitrocellulose or polyvinylidene difluoride (PVDF) membrane and appropriate molecular weight cutoff.
7. Rapid-flow vacuum filtration units with pore size 0.22 μm.
8. Thin Layer Chromatography (TLC) plates.

2.4 Isothermal Titration Calorimetry (ITC) and Small-Angle X-Ray Scattering (SAXS)

1. Sample buffer (*see* **Note 5**).
2. Gel-purified RNA that is fully exchanged into sample buffer (*see* **Note 6**).
3. Ligand solution or powder (*see* **Note 7**).
4. Centrifugal filter unit (as described in Subheading 2.3)

2.5 Atomic Force Microscopy

See Chapter 9, High-resolution atomic force microscopy imaging of RNA, Subheading 2, for a list of materials.

3 Methods

3.1 In Vitro Transcription (IVT) and RNA Purification

The procedures listed here for the preparation and purification of *Tte*AdoCbl may be different from that which is required for other samples, and experimental details may need to be modified accordingly.

1. Prepare *in vitro* transcription reactions at an appropriate scale (*e.g.*, 5 mL) in 1X IVT reaction buffer containing 50 mM Na-HEPES (pH 8.0), 25 mM $MgCl_2$, 10 mM dithiothreitol (DTT), 4 mM of each nucleotide triphosphate, 1 U/mL yeast inorganic pyrophosphatase (optional) (*see* **Note 8**), 0.05 mg/mL T7 RNA polymerase (or appropriate number of units per manufacturer's protocol), and 50–100 nM double-stranded DNA template with a T7 RNA polymerase promoter sequence.
2. Incubate the IVT reaction at 37 °C, with agitation, for 3 h.
3. Centrifuge at 4500 g for 20 min to remove the magnesium pyrophosphate precipitates.
4. Load the supernatant to a 6–10% native polyacrylamide gel of appropriate size, affixed to a vertical slab gel electrophoresis system containing 1× TBE buffer (Fig. 1a).
5. Using a high-voltage electrophoresis power supply, run the gel at 15 W, 90 V for ~16 h, or until the target RNA band has been sufficiently separated from impurities.
6. Place the resolved gel on a sheet of plastic wrap and position it over a UV fluorescent TLC plate (*see* **Note 9**).
7. Excise the target RNA transcript from the gel using a fresh razor blade and transfer the gel pieces to a sterile mortar.
8. Crush the gel pieces thoroughly using a sterile, disposable pipette or pestle, and transfer to a sterile 50 mL conical tube.
9. Add RNA elution buffer (300 mM NaOAc, pH 5.3, 0.1 mM EDTA) at a 3:1 (buffer:gel) volume ratio.
10. Elute the RNA from the gel by incubating at 4 °C, with agitation, for at least 12 h.

11. Centrifuge at 5000 g for 10 min to sediment the gel debris. Decant the eluted RNA into a fresh tube.
12. Filter the supernatant through a rapid-flow vacuum filtration unit with pore size 0.22 μm.
13. Using a centrifugal filter unit with a proper molecular weight cutoff, buffer-exchange into the desired experimental sample buffer and concentrate to a suitable stock concentration.

3.2 ITC Experiments

3.2.1 RNA Sample and Ligand Solution Preparation

1. Dialyze overnight at 4 °C an appropriate amount of stock RNA against the desired ITC binding buffer (*see* **Note 10**) to achieve complete buffer exchange prior to ITC measurements.
2. Use the dialysis buffer to dilute the RNA to the desired concentration (*e.g.*, 50 μM) and to dissolve the titrant (ligand) (*e.g.*, adenosylcobalamin (Sigma-Aldrich, United States)) at a concentration of 500 μM (*i.e.*, tenfold higher than the RNA concentration).
3. Clean the ITC injection syringe and sample cell using DEPC-water and dry completely with pure Argon gas prior to the loading of sample and titrant.
4. Fill the sample cell completely with RNA sample (200 μL for MicroCal iTC200, Malvern Panalytical Ltd, United Kingdom).
5. Fill the titration syringe with titrant (40 μL for MicroCal iTC200).
6. Set ITC data acquisition parameters, including the number of titration steps, gain sensitivity, minimum titration step volume, equilibration time between peaks, and titration syringe paddle spin rate (*see* **Note 11**).

3.2.2 ITC Data Processing and Singular Value Decomposition (SVD) Analysis

1. Perform baseline correction, subtraction, and peak integration of thermograms using ITC data processing software (*e.g.*, NITPIC [20]) (Fig. 2a).
2. Fit the data with different binding models (*e.g.*, one-site, two-site) and calculate respective thermodynamic parameters from resulting isotherms using, *e.g.*, SEDPHAT [21] (Fig. 2b).
3. Perform singular value decomposition (SVD) analysis (using *e.g.*, MATLAB and Statistics Toolbox, The MathWorks, Inc., Natick, Massachusetts, United States) on the series of titration points of the ITC thermograms. Use the series of heat compensation profiles related to each titration point to produce an m × n matrix, M, as input for SVD analysis, of which m corresponds to the number of the recorded heat compensation profiles and n corresponds to the number of titration points at different ligand:RNA ratios.

4. Based on the principal components, derive the individual isotherms that rationally reconstruct the raw ITC thermogram (Fig. 2c, d).
5. Fit the isotherms associated with each principal component with nonlinear regression fitting using different titration models (*e.g.*, one-site and two-site binding models) to determine the number states of the ligand-binding reaction, as described in [22].
6. Summarize and interpret the fitting results (thermodynamic parameters) for all SVD-derived isotherms.

3.3 Structure Calculation Using AFM Particle Topographic Restraints

Prepare AFM samples and record high-resolution images of the RNA under the desired buffer conditions (*see* **Note 12**).

3.3.1 Preparation of AFM Topographic Restraints and Initial Structure Coordinates

1. Open AFM images using conventional software (*e.g.*, Gwyddion [23], or MountainSPIP^{TM}) that can analyze scanning probe microscopy data.
2. Select and crop a single particle of interest on the AFM image.
3. Convert the particle topographic image into a text file in a three-column format with x, y, and z coordinates (in nanometers).
4. Using CafeMol, a CG biomolecular modeling and simulation software [24], generate initial coarse-grained (CG) RNA structure coordinates from any known atomic model and a native contacts file that defines hydrogen bonds, base stacking, tertiary interactions, *etc.*, associated with the model.

3.3.2 Recapitulate Topological Structure Using Dynamic Fitting Algorithm

1. Format an "input file" containing AFM topographic restraints and all required information pertaining to molecule simulation calculations, including paths, number of time steps, temperature, AFM-related settings, and parameters (*see* **Note 13**).
2. Add additional distance restraints, as necessary, by modifying the native contacts file with different weighting factors to prevent distortion of RNA structure during the dynamic fitting process.
3. Perform the calculation using the dynamic fitting module in CafeMol [25] with the previously generated input file, initial CG structural coordinates, and associated native contacts file.
4. Gradually increase the κ value to increase the AFM potential during fitting. The AFM potential, V_{AFM}, is defined as $V_{AFM}=\kappa N k_B(1\text{-}CC)$, where N is the number of atoms in the initial model, k_B is Planck's constant, κ is the scaling factor that is linearly proportional to the AFM potential, and CC is the correlation coefficient.
5. Select the representative structure candidates based on the CC score and structural energy.

3.4 SAXS Data Acquisition and Ensemble Fitting Using AFM-Derived Structures

3.4.1 SAXS Instrument Setup

1. Use of a synchrotron SAXS beamline is highly recommended. In-house SAXS systems generally have insufficient photon flux for measuring samples at extremely low concentrations required for the hybrid method.
2. SAXS experiments should be carried out in batch mode for studying the mixture of heterogeneous conformations in the ensemble.
3. Configure the instrument to cover a wide q range in a single measurement (*i.e.*, without changing the sample-detector distance). The lowest q-value, q_{min}, should be $\leq 0.3/R_g$ (radius of gyration) (*see* **Note 14**).

3.4.2 Sample Preparation for SAXS Experiments

1. Prepare the RNA sample at a minimum of three concentrations in the same sample buffer used for AFM and ITC experiments. The exact matching buffer is critical. Exchange the RNA sample into the appropriate buffer using a centrifugal filtering unit (>1000× dilution), or by dialysis overnight at 4 °C (*see* **Note 15**).
2. For ligand-containing samples, choose a ligand concentration that is appropriate based on ITC analyses. When measuring buffer only for background subtraction of these samples, the ligand concentration should best approximate the amount of free ligand in each of the RNA samples (*see* **Note 16**).

3.4.3 SAXS Data Acquisition and Processing

1. Determine a suitable exposure time and the number of frames for complete sample measurement. Exposure times should be adjusted accordingly throughout the experiment to achieve adequate signal to noise with minimal radiation damage (*see* **Note 17**).
2. Set the flow rate, shake speed, and the number of repeating cycles for the flow cell according to the total time needed for a complete sample measurement, based on the exposure time, shutter open/close time, detector readout time, and pauses between frames.
3. Collect data sets for each buffer and matching RNA sample, being sure to wash the capillary thoroughly each time before loading the next buffer into the capillary.
4. The raw SAXS data are two-dimensional scattering patterns. Reduce the raw two-dimensional SAXS patterns to a one-dimensional scattering profile using the dedicated software provided by the beamline. The reduced data contain three columns that report the scattering vectors and the mean experimental intensity and mean experimental error for each scattering vector.

5. Select reduced one-dimensional scattering profiles from proper frames by removing outliers and averaging them. The averaging should be performed separately for each individual sample or buffer.
6. Subtract buffer data from their respective RNA solution data. Use subtracted data for further analysis.
7. The procedures for basic SAXS analysis of macromolecules exist elsewhere [26], including the use of PRIMUS/qt in the ATSAS package for buffer subtraction, Guinier analysis, Kratky analysis and P(r) distribution.

3.4.4 SAXS Ensemble Fitting Guided by AFM-Derived Structures

The total solution scattering intensity from an ensemble of conformations can be synthesized as:

$$\mathrm{I}_{\mathrm{syn}}(q_i) = \sum\nolimits_{k=1}^{N_{ens}} I_{\mathrm{calc}}^{k}(q_i) \cdot \nu_k \qquad (1)$$

where ν_k and $I_{\mathrm{calc}}^{k}(q)$ are the volume fraction and calculated solution scattering intensity from the kth component/conformation, respectively. N_{ens} is the number of conformations in the ensemble, q is magnitude of the scattering vector, $q = (4\pi/\lambda)\sin\theta$, where 2θ is the scattering angle and λ is the wavelength of radiation.

1. Calculate the solution scattering intensity $I_{\mathrm{calc}}^{k}(q)$ from each atomic structure (RNA conformer) elucidated from AFM images (Subheading 3.3) using CRYSOL 3.0 [27] with q range $0 < q < 0.50$ Å^{-1}.
2. Use the volume tallies derived from AFM images (Fig. 3a) to generate initial values for the volume fractions (ν_k) for each conformer.
3. Synthesize scattering intensity curve from the ensemble of RNA conformations using Eq. 1 and compare synthesized results with experimental SAXS data (Fig. 3b) (*see* **Note 18**). The goodness-of-fit for each ensemble applied in this way can be measured by comparing the synthesized intensity scattering profile, $I_{\mathrm{syn}}(q)$, with that of the experimental data $I_{\mathrm{saxs}}(q)$.

$$\chi^2 = \frac{1}{N_q - 1} \sum\nolimits_{i=1}^{N_q} \left[\frac{I_{\mathrm{saxs}}(q_i) - \mathrm{a} \cdot I_{\mathrm{syn}}(q_i) - \mathrm{b}}{\sigma(q_i)} \right]^2$$

where $\mathrm{a} = \sum_{i=1}^{N_q} I_{\mathrm{saxs}}(q_i) \cdot I_{\mathrm{syn}}(q_i) / \sum_{i=1}^{N_q} I_{\mathrm{saxs}}(q_i) \cdot I_{\mathrm{saxs}}(q_i)$, b is background offset, N_q is number of experimental points, $\sigma(q)$ is the experimental error.

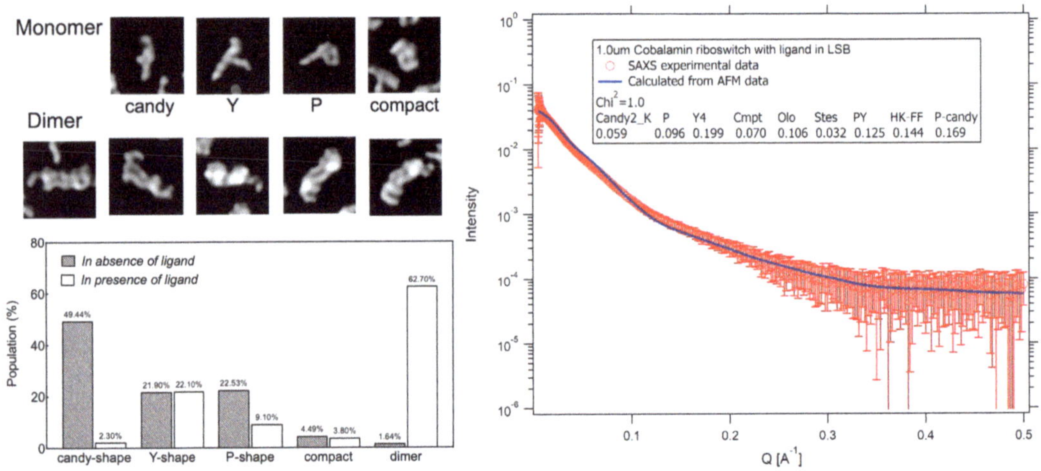

Fig. 3 AFM single particle visualization and SAXS ensemble characterization of *Tte*AdoCbl. (**a**) Various monomer and dimer conformations of *Tte*AdoCbl observed by AFM, and their populations. (**b**) Synthesized solution scattering curve (blue) using the structure ensemble defined by AFM-observed populations fit to the experimental SAXS data (red)

4 Notes

1. The materials listed here (*e.g.*, instruments, buffers, reagents) were used in the study of *Tte*AdoCbl, and may differ from what is used in other experiments.
2. DTT or other reducing agents (*e.g.*, TCEP, beta-mercaptoethanol) are crucial for the transcription activity of T7 RNA polymerase, but can be highly unstable. Addition of DTT powder directly is recommended in preparing, *e.g.*, 5X IVT reaction buffer (50 mM Na-HEPES (pH 8.0), 25 mM $MgCl_2$, 10 mM DTT), and kept at −20 °C until use.
3. T7 promoter sequence 5′-TAATACGACTCACTATAGGG-3′ was used for *Tte*AdoCbl. The third G upstream from the 3′ terminus was critical to improving transcription yield. DNA templates can be either restriction-enzyme-digested plasmid (linear DNA) or amplified PCR product after proper RNase-free cleanup procedures.
4. T7 RNA polymerase can either be purchased commercially or prepared in-house, and should be kept at −20 °C with cryo-protectant. *Tte*AdoCbl was transcribed using an in-house prepared T7 RNA polymerase.
5. The exact same sample buffer should be used for all experiments to prevent buffer mismatches of pH, reagent concentrations, *etc.*, between titrant and RNA samples, and between samples measured using different methods (*e.g.*, ITC, SAXS). Buffer mismatches can significantly convolute data analysis and interpretation.

6. Recommended RNA concentrations are 20–50 μM for ITC and 0.25–2 mg/mL for SAXS experiments.
7. Light-sensitive ligands (*e.g.*, vitamin B12) should be kept in light-resistant vials (amber glass) or opaque tubes.
8. The addition of pyrophosphatase breaks down inorganic pyrophosphate generated as a byproduct of transcription reaction cycles, and may improve yield in some cases. Typically, one unit is defined as the amount of enzyme that hydrolyzes 1 μmol of inorganic pyrophosphate in 1 min at 25 °C
9. The coated surface on the plate is excited by placing a near-UV source (254 or 365 nm) over the gel. Dark areas are observed where RNA in the gel absorbs the UV light.
10. The buffer used for *Tte*AdoCbl ITC experiments contained 50 mM MES pH 6.5, 1 mM magnesium chloride, and 10 mM potassium chloride.
11. For *Tte*AdoCbl ITC experiments, the following settings were used: 20 titration steps, high gain sensitivity, minimum titration step volume of 2 μL, equilibrium time between each peak of 180 s, titration syringe paddle spin rate of 750 rpm.
12. *See* Chapter 9, High-resolution atomic force microscopy imaging of RNA, Subheading 3, for detailed procedures for imaging RNA by high-speed AFM.
13. Refer to the CafeMol user manual for additional information on preparing the input file and related procedures [24].
14. For *Tte*AdoCbl SAXS experiments, the q range was $0.005 < q < 0.88$ Å−1 and data were fit up to 0.5 $Å^{-1}$.
15. Ideally, the exact same sample conditions should be used for both SAXS and AFM experiments. However, the RNA concentrations used for AFM imaging (nM) are much too low for SAXS measurements. The RNA concentration range, however, should include the minimum concentration at which a reasonable SAXS signal can be detected, which will vary by beamline.
16. For example, in a *Tte*AdoCbl SAXS experiment, the 2 μM RNA sample contained 20 μM ligand. Assuming a 1:1 binding ratio, then, a buffer containing 18 μM ligand was used for the background subtraction for this sample.
17. The optimal exposure time will vary by sample, sample concentration, and beamline. For *Tte*AdoCbl SAXS experiments, conducted at APS 12-ID-B (Argonne National Lab), data were acquired at 1 s exposure per frame and 225 frames for each sample.
18. For *Tte*AdoCbl, SAXS data acquired at a concentration of 0.07 mg/mL (1 μM) were selected for ensemble fitting.

Acknowledgment

This research was supported by the Intramural Research Program of the NIH, National Cancer Institute, Center for Cancer Research (Y-X. Wang).

References

1. Svergun DI, Koch MHJ (2003) Small-angle scattering studies of biological macromolecules in solution. Rep Prog Phys 66:1735–1782
2. Jeffries CM, Ilavsky J, Martel A et al (2021) Small-angle X-ray and neutron scattering. Nat Rev Methods Primers 1:70
3. Lipfert J, Doniach S (2007) Small-angle X-ray scattering from RNA, proteins, and protein complexes. Annu Rev Biophys Biomol Struct 36:307–327
4. Brosey CA, Tainer JA (2019) Evolving SAXS versatility: solution X-ray scattering for macromolecular architecture, functional landscapes, and integrative structural biology. Curr Opin Struct Biol 58:197–213
5. Chen Y, Pollack L (2016) SAXS studies of RNA: structures, dynamics, and interactions with partners. Wiley Interdiscip Rev RNA 7: 512–526
6. O'carroll IP, Fan L, Kroupa T et al (2020) Structural mimicry drives HIV-1 rev-mediated HERV-K expression. J Mol Biol 432:166711
7. Fang X, Wang J, O'carroll IP et al (2013) An unusual topological structure of the HIV-1 Rev response element. Cell 155:594–605
8. Bofill-De Ros X, Kasprzak WK, Bhandari Y et al (2019) Structural differences between Pri-miRNA paralogs promote alternative drosha cleavage and expand target repertoires. Cell Rep 26:447
9. Wang YX, Zuo XB, Wang JB et al (2010) Rapid global structure determination of large RNA and RNA complexes using NMR and small-angle X-ray scattering. Methods 52:180–191
10. Harding RJ, Loppnau P, Ackloo S et al (2019) Design and characterization of mutant and wildtype huntingtin proteins produced from a toolkit of scalable eukaryotic expression systems. J Biol Chem 294:6986–7001
11. Cantara WA, Olson ED, Musier-Forsyth K (2017) Analysis of RNA structure using small-angle X-ray scattering. Methods 113:46–55
12. Da Vela S, Svergun DI (2020) Methods, development and applications of small-angle X-ray scattering to characterize biological macromolecules in solution. Curr Res Struct Biol 2:164–170
13. Song H, Ji XH (2019) The mechanism of RNA duplex recognition and unwinding by DEAD-box helicase DDX3X. Nat Commun 10:3085
14. Johnson JE Jr, Reyes FE, Polaski JT et al (2012) B12 cofactors directly stabilize an mRNA regulatory switch. Nature 492:133–137
15. Schneidman-Duhovny D, Hammel M, Tainer JA et al (2016) FoXS, FoXSDock and Multi-FoXS: single-state and multi-state structural modeling of proteins and their complexes based on SAXS profiles. Nucleic Acids Res 44: W424–W429
16. Cantero-Camacho A, Fan LX, Wang YX et al (2017) Three-dimensional structure of the 3′ X-tail of hepatitis C virus RNA in monomeric and dimeric states. RNA 23:1465–1476
17. Petoukhov MV, Billas IML, Takacs M et al (2013) Reconstruction of quaternary structure from X-ray scattering by equilibrium mixtures of biological macromolecules. Biochemistry 52:6844–6855
18. Micsonai A, Wien F, Kernya L et al (2015) Accurate secondary structure prediction and fold recognition for circular dichroism spectroscopy. Proc Natl Acad Sci USA 112: E3095–E3103
19. Schmidt M, Rajagopal S, Ren Z et al (2003) Application of singular value decomposition to the analysis of time-resolved macromolecular x-ray data. Biophys J 84:2112–2129
20. Keller S, Vargas C, Zhao H et al (2012) High-precision isothermal titration calorimetry with automated peak-shape analysis. Anal Chem 84: 5066–5073
21. Zhao H, Piszczek G, Schuck P (2015) SEDPHAT – a platform for global ITC analysis and global multi-method analysis of molecular interactions. Methods 76:137–148
22. Poon GM (2010) Explicit formulation of titration models for isothermal titration calorimetry. Anal Biochem 400:229–236
23. David Nečas PK (2012) Gwyddion: an open-source software for SPM data analysis. CEJPh 10:181–188
24. Kenzaki H, Koga N, Hori N et al (2011) Cafe-Mol: a coarse-grained biomolecular simulator

for simulating proteins at work. J Chem Theory Comput 7:1979–1989

25. Niina T, Fuchigami S, Takada S (2020) Flexible fitting of biomolecular structures to atomic force microscopy images via biased molecular simulations. J Chem Theory Comput 16: 1349–1358
26. Mrozowich T, Mclennan S, Overduin M et al (2018) Structural studies of macromolecules in solution using small angle X-Ray scattering. J Vis Exp (141):e58538
27. Franke D, Petoukhov MV, Konarev PV et al (2017) ATSAS 2.8: a comprehensive data analysis suite for small-angle scattering from macromolecular solutions. J Appl Crystallogr 50: 1212–1225

Chapter 12

Near-Atomic Resolution Cryo-EM Image Reconstruction of RNA

Shanshan Li, Kaiming Zhang, and Wah Chiu

Abstract

The rapid development of cryogenic electron microscopy (cryo-EM) enables the structure determination of macromolecules without the need for crystallization. Protein, protein–lipid, and protein–nucleic acid complexes can now be routinely resolved by cryo-EM single-particle analysis (SPA) to near-atomic or atomic resolution. Here we describe the structure determination of pure RNAs by SPA, from cryo-specimen preparation to data collection and 3D reconstruction. This protocol is useful to yield many cryo-EM structures of RNA, here exemplified by the *Tetrahymena* L-21 ScaI ribozyme at 3.1-Å resolution.

Key words Single-particle analysis, RNA, Ribozyme, Protocol, Cryo-EM

1 Introduction

Cryogenic electron microscopy (cryo-EM) and tomography (cryo-ET) have become routine and popular techniques for structure determination of biological samples, from biochemically purified biomolecules to whole cells, yielding unprecedented details and elucidating functional mechanisms [1–4]. Such advances are attributable to the availability of direct electron detectors [5, 6], automated high-throughput data collection electron microscopes [7], and freely accessible data processing software [8–10]. Cryo-EM single-particle analysis (SPA) is now capable of competing with X-ray crystallography and NMR spectroscopy to resolve macromolecular structures at near-atomic or atomic resolution without the need for crystallization [1, 4]; it also has the unique advantage of resolving multiple structures of samples with high heterogeneity or flexibility [11–14].

While protein-coding genes only occupy a very small portion of the human genome, the vast majority (~97%) of transcriptional products are non-protein-coding RNAs, including transfer RNA (tRNA) and ribosomal RNA (rRNA) involved in mRNA

Jienyu Ding et al. (eds.), *RNA Structure and Dynamics*, Methods in Molecular Biology, vol. 2568,
https://doi.org/10.1007/978-1-0716-2687-0_12,

translation, and noncoding RNAs involved in gene regulation, RNA processing, etc. They exert their molecular functions by folding into complex three-dimensional (3D) structures. The number of 3D structures of pure RNAs is far less than that of proteins, limiting our understanding of RNA structure–function relationships. SPA has been rarely utilized in the structural characterization of pure RNAs, due to the perception of being difficult with their small size and potentially inherent flexibilities. Recently, we have taken advantage of the advances in cryo-EM to determine the structures of a broad spectrum of RNAs ranging from 88 to 400 nts at sub-nanometer to near-atomic resolution [15–17]. In the best case, *Tetrahymena* L-21 ScaI ribozyme (388 nts) can be determined to 3.1 Å resolution [18].

2 Materials

2.1 Reagents

1. All chemical reagents were purchased from Thermo Fisher Scientific unless otherwise indicated.
2. Milli-Q water.
3. Liquid N_2 (*see* **Note 1**).
4. High-purity ethane (*see* **Note 2**).

2.2 Equipment

1. Surgical masks.
2. Felt-tip pen.
3. RNase-free pipettes and pipette tips.
4. 1.5-ml RNase-free conical tubes (LABSELECT, cat. no. MCT-001-150).
5. AMPure XP Beads (Beckman Coulter, cat. no. A63881)..
6. RNA Clean and Concentrator (Zymo Research, cat. no. R1015).
7. Vitrobot filter paper Ø55/20 mm, Grade 595 (TED PELLA, cat. no. 47000-100).
8. PELCO easiGlow Glow Discharge System (TED PELLA, cat. no. 91000).
9. Vitrobot Mark IV Plunge-freezer (Thermo Fisher Scientific).
10. Quantifoil EM grids (Electron Microscopy Sciences).
11. Tweezers for EM grids (TED PELLA).
12. Grid boxes (TED PELLA, cat. No. 160-42).
13. Gripper tool for grid Box (TED PELLA, cat. no. 160-48).
14. AutoGrid (Thermo Fisher Scientific).
15. C-ring (Thermo Fisher Scientific).
16. Clipping tool for C-ring (Thermo Fisher Scientific).

17. Grid clipping station (Thermo Fisher Scientific).
18. AutoGrid boxes (Thermo Fisher Scientific).
19. AutoGrid loading station (Thermo Fisher Scientific).
20. Titan Krios G3i cryo-transmission electron microscope (Thermo Fisher Scientific) with GIF energy filter (Gatan).
21. Gatan K2 or K3 Summit direct electron detector (AMETEK).

2.3 Software

1. EPU (Thermo Fisher Scientific).
2. MotionCor2 https://emcore.ucsf.edu/ucsf-software.
3. CTFFIND4 https://grigoriefflab.umassmed.edu/ctffind4.
4. EMAN2 https://blake.bcm.edu/emanwiki/EMAN2.
5. Relion https://www2.mrc-lmb.cam.ac.uk/relion.
6. cryoSPARC https://cryosparc.com/.
7. Chimera http://www.cgl.ucsf.edu/chimera.
8. Coot http://www2.mrc-lmb.cam.ac.uk/personal/pemsley/coot/.
9. Phenix https://www.phenix-online.org.

3 Experimental and Computational Methods

Herein we describe the structure determination of pure RNA specimens using SPA, from sample preparation to data collection and 3D reconstruction (Fig. 1).

3.1 RNA Sample Preparation

DNA templates were assembled by polymerase chain reaction (PCR) of oligonucleotides designed with Primerize [19, 20] and purchased from Integrated DNA Technologies. These templates were purified by AMPure XP Beads according to the manufacturer's instructions. RNA was then in vitro transcribed from these DNA templates and purified using RNA Clean and Concentrator following the manufacturer's instructions.

3.2 RNA Folding

About 20 μl of 25 μM RNAs in 50 mM Na-HEPES, pH 8.0 were incubated at 90°C for 3 min on a thermal cycle machine (Bio-Rad), then cooled at room temperature for 10 min. Subsequently, $MgCl_2$ was added to the final concentration of 10 mM, incubated at 50°C for 30 min, and cooled at room temperature for 10 min.

3.3 Cryo-EM Freezing and Data Collection

1. Glow discharge grids (200-mesh R2/1 Quantifoil Copper grids): 35 s using PELCO easiGlow at a plasma current of 15 mA.
2. Immediately freeze the RNA samples after folding using a standard protocol for the commercial Vitrobot Mark IV

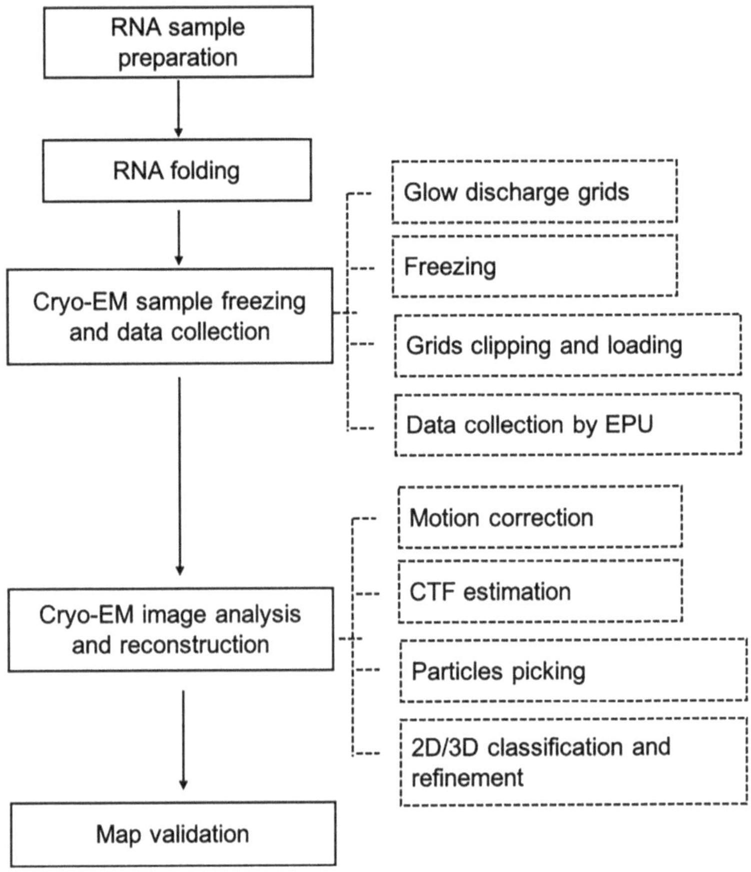

Fig. 1 Schematic of the workflow of RNA 3D structure determination by cryo-EM

plunge-freezer. The relevant parameters were optimized as follows: 3 μl of L21 ScaI ribozyme RNAs (25 μM); blotting time 2–4 s; wait time 5 s; 4°C temperature; ~100% humidity (*see* **Note 3**). The blotting paper was standard Vitrobot filter paper Ø55/20 mm, Grade 595.

3. Sample clipping and loading to the autoloader of Titan Krios microscope.
4. Collect the data with the EPU software following the manufacturer's instructions (*see* **Note 4**). The relevant parameters were optimized as follows: a magnification of ×105,000 (corresponding to a calibrated sampling of 0.82 Å per pixel); 30 individual frames per image; an exposure time of 2.5 s; a dose rate of 22.9 electrons per second per Å^2; a defocus range of −1.0 to −2.5 μm (*see* **Notes 5** and **6**).

3.4 Cryo-EM Image Analysis and Reconstruction

Data processing was performed using a series of steps with different software (Fig. 2) (*see* **Note 7**). Unless otherwise stated, the software was used according to the instructions for each software.

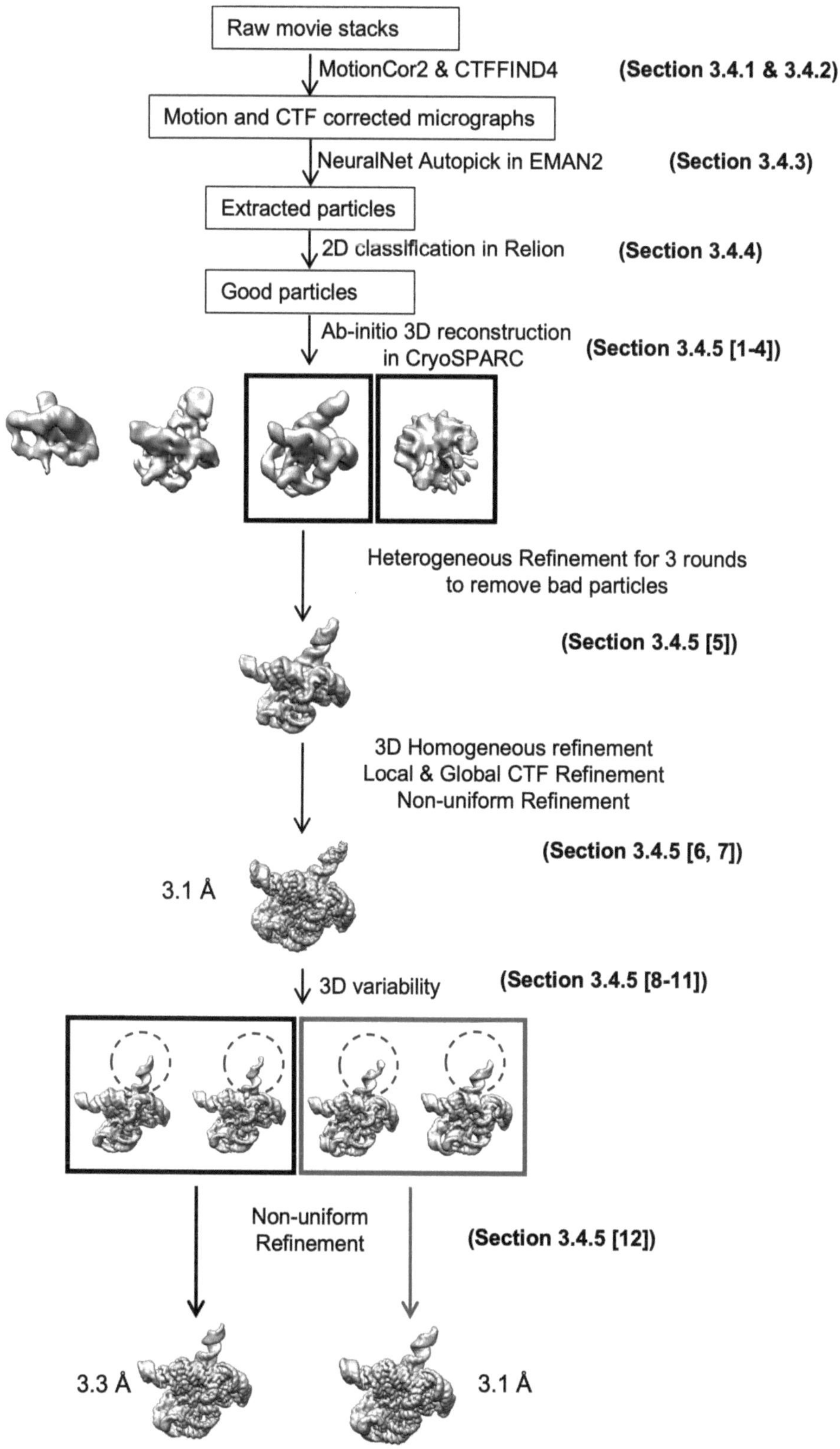

Fig. 2 Workflow of single-particle cryo-EM analysis of *Tetrahymena* L-21 ScaI ribozyme

3.4.1 All Micrographs Were Motion-Corrected Using MotionCor2

Beam-induced sample motion blurs the captured micrographs and limits the final resolution. Apply MotionCor2 [5], a software tool for anisotropic correction of beam-induced motion, and select micrographs with less motion for further use. The micrographs were motion-corrected with "B-factor" of 150, "number of patches X, Y" of 5, 5, and "group frames" of 5, while using default values for all other parameters. Since MotionCor2 is embedded as a wrapper in Relion [10], one can sort the micrographs.star by "rlnMotionEarly" using the "subset selection" option in Relion and select the micrographs with less motion (*e.g.*, "rlnMotionEarly" <10).

3.4.2 Contrast Transfer Function (CTF) Was Determined Using CTFFIND4

After the CTF fits, sort the micrographs according to resolutions estimated by CTFFIND4 [21] and remove the micrographs with a maximum detectable CTF signal no better than 5 Å. The CTF estimation was done on the micrographs without dose-weighting in order to better fit the low resolution signal. CTFFIND4 is accessible in Relion. One can sort the quality of the micrographs using micrographs_ctf.star based on "rlnCtfMaxResolution" with the "subset selection" option and select the micrographs with "rlnCtfMaxResolution" better than 5 Å for further processing.

3.4.3 All Particles Were Autopicked Using the NeuralNet Option in EMAN2

(*see* **Notes 8** and **9**)

1. Launch EMAN2 [22] and input all the micrographs.
2. Use "Filter display" (red box 1 in Fig. 3) to increase contrast to make particles more visible.
3. Set up three networks: Good References, Bad References, and Background References (red box 2–4 in Fig. 3). Use a box size that is 1.5–2 times the particle size and divisible by 2. Select ~50 good particles from multiple micrographs at various defocuses to Good References; select ~30 bad particles like ice contamination or large aggregation from multiple micrographs to Bad References; select ~30 pure noise regions from multiple micrographs to Background References.
4. Train the three networks (red box 5 in Fig. 3). Autobox one image (red box 6 in Fig. 3). Try out different threshold values (red box 7 in Fig. 3) to get the maximum number of good particles. When an optimal value is found, click "Autobox all" to pick particles for all the micrographs (red box 8 in Fig. 3).
5. Manually check all particles and remove the bad particles including carbon edge and big ice in each micrograph.
6. Generate particle coordinates and import them to Relion [10].

3.4.4 2D Classification in Relion

Perform two rounds of 2D classification in Relion to remove 2D class averages with less resolved features (*see* **Notes 10** and **11**).

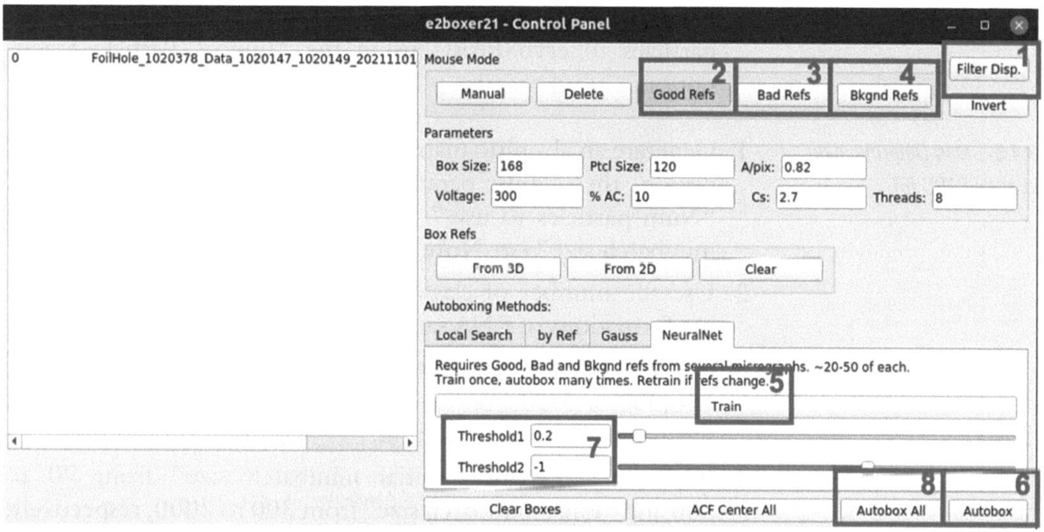

Fig. 3 NeuralNet particles autopicking in EMAN2

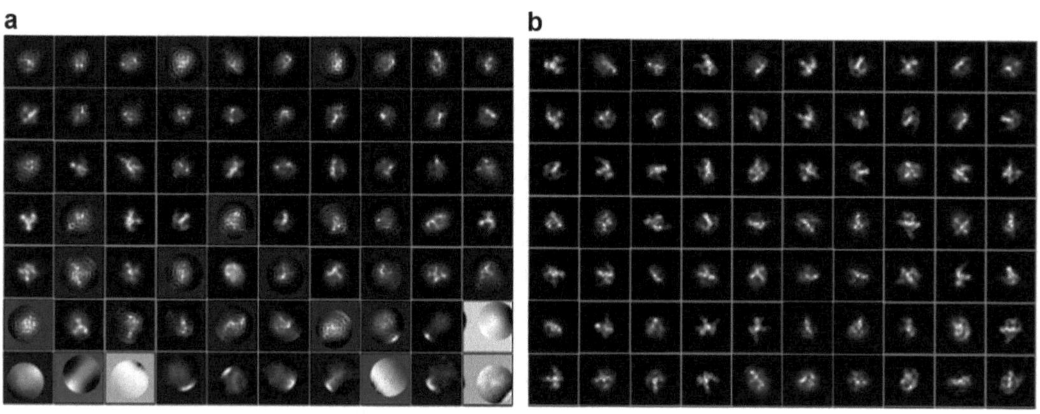

Fig. 4 2D classification of the extracted particles. (**a**) First round. (**b**) Second round. The particles from the classes with red boxes were selected

Each round is an iterative process and we use the default of 25 iterations (*see* **Note 12**)

1. First round (preliminary screening): set "number of classes" to ~80, "limit resolution E-step" to 16 Å to prevent overfitting. The purpose of the first round is to remove 2D class averages that have ice contamination and a very noisy background (Fig. 4a).
2. Second round (fine screening): set "number of classes" to ~80, "limit resolution E-step" to 12 Å. The purpose of this round is to select 2D class averages with helical features (Fig. 4b).

3. Execute "Particle Extraction" in Relion, and import these particles to cryoSPARC using the "Import Particle Stack" option.

3.4.5 Map Determination in cryoSPARC [8]

1. Generate an ab-initio map: use the "Ab-initio Reconstruction" option, then adjust parameters including "Num of classes," "Num particles to use," "Initial minibatch size," and "Final minibatch size" (*see* **Note 13**).
2. Use the number of classes of 4 to initially identify multiple conformations of RNA samples as well as to find junk particles.
3. (Optional) Set "Num particles to use" to a small number (*e.g.*, 100,000 or 5-10% of the entire dataset) if the dataset is large, speeding up the computation.
4. (Optional) Increase "Initial minibatch size" from 90 to 300 and "Final minibatch size" from 300 to 2000, respectively, to increase signal-to-noise ratio and improve the reliability of the initial map production for low-molecular-weight samples.
5. Perform multiple rounds of Heterogeneous Refinement to reclassify particles to further remove junk particles.

   Select the particles from the good 3D maps from the previous "Ab-Initio Reconstruction" task. A good 3D map should have RNA features, like the major groove and double-stranded helices. Select one good map and one bad map from the previous "Ab-Initio Reconstruction" task as the initial models (Fig. 2) to sort the particles into two groups. Repeat this step using the particles from the good map for another 2–3 rounds.
6. Homogenous and CTF Refinement.

   Refine the orientation parameters of the particles yielding the good 3D map to an improved map with higher resolution using Homogenous Refinement. The map resolution is defined by the gold-standard Fourier shell correlation (FSC). The CTF parameters were determined initially from each micrograph from Subheading 3.4.2. However, there could be local variations of the CTF due to nonuniform ice thickness, and misalignment or aberrations in the microscope itself. The options of Local CTF refinement (per-particle defocus) and Global CTF Refinement (per-exposure-group beam-tilt, trefoil, spherical aberration, tetrafoil, and anisotropic magnification) can be used to improve the CTF corrections per particle. Such refinement can be coupled with the Homogeneous Refinement to achieve an optimal map(s) with the most resolved densities (Fig. 2) (*see* **Note 14**).
7. Nonuniform Refinement.

   Because different regions of a density map may be resolved differently due to local flexibility, we can refine a single good

3D map to better resolution and map quality. In this refinement step, we often include the Local and Global CTF refinement. In most cases, Nonuniform Refinement improves resolution by 0.1–0.3 Å over Homogeneous Refinement as done in the previous step.

8. 3D classification without particle orientation realignment using "3D Variability."

   The previous steps refine the data based on orientation refinement of 2D particle images while the next stage of map refinement is to sort out particle heterogeneity based on 3D classification of local regions of the structure without changing the particle orientations.

9. Input particles and mask from the previous Nonuniform Refinement task; set the "Filter Resolution" to ~6 Å in "3D Variability."
10. Display the structural variability in different cluster classes for classification by choosing "output mode" to cluster, "Number of clusters" to 4, "Filter resolution" to ~6 Å.
11. Display the 3D map variability to see the motion among the components by using a linear "movie" mode. Set "output mode" to simple, "Number of frames" to 20, "Filter resolution" to ~6 Å.
12. Merging different classes after 3D classification and performing another round of Nonuniform Refinement.

    Combine 3D classes with similar features and very minor conformational changes, and do the Nonuniform Refinement individually (Fig. 2).

3.5 Map Sharpening and Masking

There are multiple maps generated in the above image processing workflow (Subheading 3.4). Each of the steps is aimed at obtaining map(s) with the best resolved densities, at least in part of the RNA molecules. The maps are accompanied with resolution estimation based on the FSC of two independently determined maps from two sets of particles (Fig. 5a). Since the map reconstruction generally suffers from a resolution deterioration due to a combination of factors from specimen integrity, image quality to data processing errors, a step of using sharpening to rescale the Fourier components at higher frequency without generating noise peaks would enhance the image density contrast. In addition, masking the density would assist the interpretation of the map. Both sharpening and masking can easily be overdone and caution needs to be exercised. Both of these operations are built automatically into the Heterogeneous, Homogeneous, and Nonuniform Refinement steps (*see* **Note 15**).

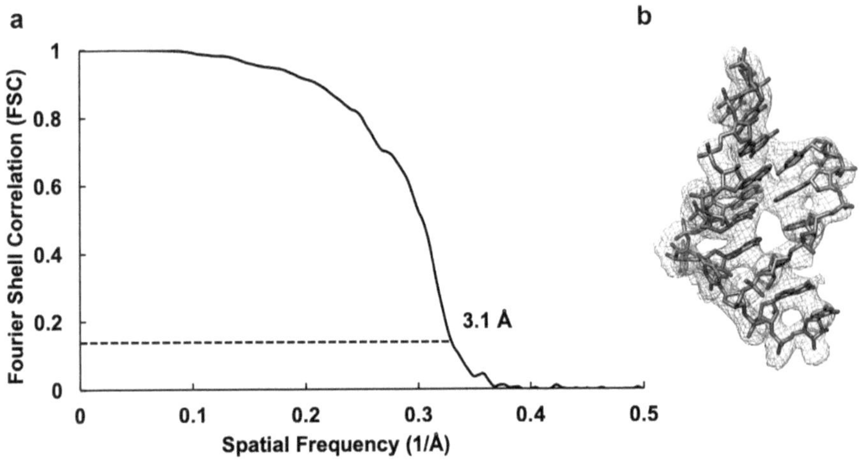

Fig. 5 3D reconstruction of *Tetrahymena* L-21 ScaI ribozyme at 3.1-Å resolution. (**a**) Gold Standard FSC curve shows 3.1-Å resolution according to the 0.143 cutoff. (**b**) A selected region to show base pairing and stacking (U267-U273 & A304-A311)

4 Map Validation

A cryo-EM map can reveal different features depending on the resolution. For a near-atomic resolution map, one can build an atomic model based on the map density, as in the case of this 3.1 Å ribozyme map, using standard crystallographic modeling tools like Coot [23] and Phenix [24]. The final model needs to follow the stereochemistry of nucleotides and concurrently fit the cryo-EM density. Base pairing and base stacking are apparent for a map at ~3.0 Å resolution (Fig. 5b). A quantitative measure of the resolvability of cryo-EM density has been introduced by using the Q-scores[25, 26]. In the event that the cryo-EM map is determined at sub-nanometer resolution, its structural interpretation can be substantiated by tagging with an additional RNA having a known structure [15]. Alternatively, the cryo-EM map can be fitted to the known structure of a certain domain of the RNA [27, 28].

5 Map and Model Submission to wwPDB

The final maps along with the metadata and models should be deposited to wwPDB using "wwPDB OneDep System" (https://deposit-2.wwpdb.org/deposition/). A wwPDB EM Validation Report will be generated to describe the "quality" of the structure. Rigorous validation of the structural features of RNA at sub-nanometer and near-atomic resolution is still an active research as cryo-EM becomes increasingly useful for such endeavors.

6 Notes

1. Use a liquid nitrogen container free of residual water. Water contamination must be avoided during any step of manipulation of the frozen grids. Besides, to avoid gasified N_2-caused asphyxiation, appropriate approaches should be used, including adequate ventilation and the use of personal protective equipment.
2. Ethane is flammable. Avoid sparks and ensure adequate ventilation.
3. Freezing parameters need to be optimized for different samples, *e.g.*, blot time, blot force, wait time, and humidity.
4. Data acquisition can be performed using different automation software, *e.g.*, EPU for the microscopes from Thermo Fisher Scientific, JADAS for the microscopes from JEOL, Latitude for the microscopes with Gatan K2/K3 detectors, and SerialEM [29] for most microscopes.
5. If the RNA specimen is smaller than 150 nt, a defocus range of -3 to -4 μm can be used for initial small dataset collection, where the high contrast of particles provides a better chance to get a good initial model.
6. Apply phase plate technique. Similar to the purpose in **Note 5**. The technique can improve the particles' contrast, which is useful to generate the initial model [17]. However, the Volta phase plate from Thermo Fisher Scientific, and the Hole-free phase plate from JEOL are not included in the factory configuration; a separate purchase is needed.
7. The image processing protocol in this chapter is an experience-based approach, but not the only one. Relion can also be used for map determination, which can be found in our previous work [18]. Alternatively, all steps of image processing can be performed in cryoSPARC only as follows: (i) Micrographs were motion-corrected using "Patch Motion Correction." (ii) CTF was determined using "Patch CTF Estimation." (iii) Particle picking: (a) Manually pick ~ 200 particles. (b) Run 2D classification to get 4–5 fair classes as templates. (c) Execute "Template Picker" to pick the particles from all micrographs. (d) Extract the particles by "Extract from micrographs." (iv) 2D Classification. The following steps are the same as the procedures described in this chapter.
8. When doing particle picking in EMAN2, make sure to center all good particles, which should also cover as many different views of the particle as possible.
9. If the RNA has a published structure or a homologous structure, one can also use that structure as the template to auto-

pick the particles in EMAN2 using the "by Ref" option. To avoid "Einstein face" bias during the particle picking [30], it is necessary to bandpass the structure to lower resolution.

10. If the goal is to obtain high-resolution 3D structures, RNA specimens that do not have high-resolution characteristics (*e.g.*, helical groove features) on the 2D class averages can be returned to sample preparation optimization without the need for further data processing.
11. When doing particle extraction, downscaling the dataset by 3-4 times for the 2D classification can speed up the data processing.
12. When running 2D classification, the default value of iteration in Relion is 25. For smaller RNAs, one can increase the number of iterations to 50-70, which may give better converged results.
13. Besides the "Ab-initio Reconstruction" option in cryoSPARC, EMAN2 can be also used for the initial model building. The steps are as follows: i) Select ~10,000 particles. ii) Run reference-free 2D classification. iii) Select good classes to build the initial model using the "Make initial model" option.
14. Local and Global CTF Refinement can be used as separate tasks. One can try different strategies to get better 3D maps. i) 3D Refinement (Either Homogeneous or Nonuniform Refinement) coupled with Local and Global CTF Refinement on the fly. ii) Run 3D Refinement, Local CTF Refinement, Global CTF Refinement, and 3D Refinement in separate and sequential tasks. iii) Run 3D refinement, Global CTF Refinement, Local CTF Refinement, and 3D refinement in separate and sequential tasks.
15. Map sharpening sometimes is not ideal from the 3D refinement job. One can do the map sharpening with the "sharpening tools" option in cryoSPARC or sharpen the map using two half maps in other software, *e.g.*, the "auto-sharpening" option in Phenix.

Acknowledgment

This work is support by National Institutes of Health grants (P41GM103832 and R21AI145647 to WC) and start-up funding by the University of Science and Technology of China (KY9100000032 and KJ2070000080 to KZ). We thank Dr. Michael F. Schmid for helpful comments on the manuscript.

References

1. Zhang K, Pintilie GD, Li S et al (2020) Resolving individual atoms of protein complex by cryo-electron microscopy. Cell Res 30: 1136–1139
2. Zhang K, Li S, Pintilie G, et al (2020) A 3.4-Å cryo-electron microscopy structure of the human coronavirus spike trimer computationally derived from vitrified NL63 virus particles. QRB Discov 1:e11
3. Roh S-H, Shekhar M, Pintilie G et al (2020) Cryo-EM and MD infer water-mediated proton transport and autoinhibition mechanisms of V complex. Sci Adv 6
4. Cogan DP, Zhang K, Li X et al (2021) Mapping the catalytic conformations of an assembly-line polyketide synthase module. Science 374:729–734
5. Zheng SQ, Palovcak E, Armache J-P et al (2017) MotionCor2: anisotropic correction of beam-induced motion for improved cryo-electron microscopy. Nat Methods 14: 331–332
6. Li X, Mooney P, Zheng S et al (2013) Electron counting and beam-induced motion correction enable near-atomic-resolution single-particle cryo-EM. Nat Methods 10:584–590
7. Cheng A, Tan YZ, Dandey VP et al (2016) Strategies for automated CryoEM data collection using direct detectors. Methods Enzymol 579:87–102
8. Punjani A, Rubinstein JL, Fleet DJ et al (2017) cryoSPARC: algorithms for rapid unsupervised cryo-EM structure determination. Nat Methods 14:290–296
9. Chen M, Bell JM, Shi X et al (2019) A complete data processing workflow for cryo-ET and subtomogram averaging. Nat Methods 16: 1161–1168
10. Scheres SHW (2012) RELION: implementation of a Bayesian approach to cryo-EM structure determination. J Struct Biol 180:519–530
11. Scheres SHW (2016) Processing of structurally heterogeneous Cryo-EM data in RELION. Methods Enzymol 579:125–157
12. Zhao Y, Schmid MF, Frydman J et al (2021) CryoEM reveals the stochastic nature of individual ATP binding events in a group II chaperonin. Nat Commun 12:4754
13. Zhang K, Zhang H, Li S et al (2019) Cryo-EM structures of vacuolating cytotoxin A oligomeric assemblies at near-atomic resolution. Proc Natl Acad Sci USA 116:6800–6805
14. Punjani A, Fleet DJ (2021) 3D variability analysis: Resolving continuous flexibility and discrete heterogeneity from single particle cryo-EM. J Struct Biol 213:107702
15. Zhang K, Zheludev IN, Hagey RJ et al (2021) Cryo-EM and antisense targeting of the 28-kDa frameshift stimulation element from the SARS-CoV-2 RNA genome. Nat Struct Mol Biol 28:747–754
16. Kappel K, Zhang K, Su Z et al (2020) Accelerated cryo-EM-guided determination of three-dimensional RNA-only structures. Nat Methods 17:699–707
17. Zhang K, Li S, Kappel K et al (2019) Cryo-EM structure of a 40 kDa SAM-IV riboswitch RNA at 3.7 Å resolution. Nat Commun 10:5511
18. Su Z, Zhang K, Kappel K et al (2021) Cryo-EM structures of full-length Tetrahymena ribozyme at 3.1 Å resolution. Nature 596: 603–607
19. Tian S, Das R (2017) Primerize-2D: automated primer design for RNA multidimensional chemical mapping. Bioinformatics 33: 1405–1406
20. Tian S, Yesselman JD, Cordero P et al (2015) Primerize: automated primer assembly for transcribing non-coding RNA domains. Nucleic Acids Res 43:522–526
21. Rohou A, Grigorieff N (2015) CTFFIND4: Fast and accurate defocus estimation from electron micrographs. J Struct Biol 192:216–221
22. Tang G, Peng L, Baldwin PR et al (2007) EMAN2: an extensible image processing suite for electron microscopy. J Struct Biol 157: 38–46
23. Emsley P, Lohkamp B, Scott WG et al (2010) Features and development of Coot. Acta Crystallogr D Biol Crystallogr 66:486–501
24. Adams PD, Afonine PV, Bunkóczi G et al (2010) PHENIX: a comprehensive Python-based system for macromolecular structure solution. Acta Crystallogr D Biol Crystallogr 66:213–221
25. Pintilie G, Zhang K, Su Z et al (2020) Measurement of atom resolvability in cryo-EM maps with Q-scores. Nat Methods 17:328–334
26. Lawson CL, Kryshtafovych A, Adams PD et al (2021) Cryo-EM model validation recommendations based on outcomes of the 2019 EMDataResource challenge. Nat Methods 18: 156–164

27. Torabi S-F, Chen Y-L, Zhang K et al (2021) Structural analyses of an RNA stability element interacting with poly(A). Proc Natl Acad Sci USA 118
28. Zhang K, Keane SC, Su Z et al (2018) Structure of the 30 kDa HIV-1 RNA dimerization signal by a hybrid Cryo-EM, NMR, and molecular dynamics approach. Structure 26: 490–498. e3
29. Mastronarde DN (2005) Automated electron microscope tomography using robust prediction of specimen movements. J Struct Biol 152:36–51
30. Henderson R (2013) Avoiding the pitfalls of single particle cryo-electron microscopy: Einstein from noise

Chapter 13

Auto-DRRAFTER: Automated RNA Modeling Based on Cryo-EM Density

Haiyun Ma, Phillip Pham, Bingnan Luo, Ramya Rangan, Kalli Kappel, Zhaoming Su, and Rhiju Das

Abstract

RNA three-dimensional structures provide rich and vital information for understanding their functions. Recent advances in cryogenic electron microscopy (cryo-EM) allow structure determination of RNAs and ribonucleoprotein (RNP) complexes. However, limited global and local resolutions of RNA cryo-EM maps pose great challenges in tracing RNA coordinates. The Rosetta-based "auto-DRRAFTER" method builds RNA models into moderate-resolution RNA cryo-EM density as part of the Ribosolve pipeline. Here, we describe a step-by-step protocol for auto-DRRAFTER using a glycine riboswitch from *Fusobacterium nucleatum* as an example. Successful implementation of this protocol allows automated RNA modeling into RNA cryo-EM density, accelerating our understanding of RNA structure–function relationships. Input and output files are being made available at https://github.com/auto-DRRAFTER/springer-chapter.

Key words RNA structure, Cryo-EM, Auto-DRRAFTER, Rosetta computational modeling

1 Introduction

RNAs can fold into intricate three-dimensional (3D) structures to perform diverse biological functions in the presence or absence of proteins [1, 2]. Single particle cryogenic electron microscopy (cryo-EM) is routinely used now to determine RNA and ribonucleoprotein (RNP) structures at near-atomic resolution [3–5]. However, limited resolution in flexible global and local regions of RNAs prevents accurate RNA modeling and poses challenges in elucidating the RNA structure–function relationship [6, 7].

Computational modeling and de novo design of RNA structures is in early stages compared to protein modeling, but continues to improve due to expanding experimental methods using X-ray crystallography [8], NMR [9], small-angle X-ray scattering (SAXS) [10], cryo-EM [11], and development of RNA modeling software

Jienyu Ding et al. (eds.), *RNA Structure and Dynamics*, Methods in Molecular Biology, vol. 2568,
https://doi.org/10.1007/978-1-0716-2687-0_13,

including Assemble2 [12], MC-sym [13], RACER [14], Phenix [15], Rosetta [16], VfoldLA [17], RNAComposer [18], NAST [19], SimRNA [20], and iFoldRNA [21]. Driving progress, the RNA-Puzzles are held routinely to provide rigorous evaluations on different aspects of existing RNA modeling software, including homology modeling, fragment assembly, and de novo prediction [22]. Recent advances in the application of machine learning algorithms to RNA modeling may also improve the accuracy of de novo RNA structure prediction [23, 24], though a scarcity of RNA 3D structures compared to protein structures may delay dramatic advances from machine learning methods, which depend on large data sets.

Development of single particle cryo-EM has allowed determination of RNA and RNP complexes at increasing throughput. However, in work so far, the majority of RNA structures are resolved at moderate to low resolution (~ 4–10 Å), which limits the use of currently available de novo modeling algorithms based on cryo-EM density at near-atomic (better than 4 Å) resolution [5, 16] and also makes manual tracing of RNA coordinates tedious or even intractable. We previously developed DRRAFTER to build RNA coordinates de novo in biologically important RNP complexes, in which RNAs were resolved at relatively low resolution [25]. DRRAFTER contains an initial manual setup that requires users' knowledge and experience in RNA modeling, which may crucially affect the performance and final results. In the recent Ribosolve pipeline, we developed an updated version of DRRAFTER, named auto-DRRAFTER, to automatically build RNA models into moderate-resolution cryo-EM maps [11]. The input to auto-DRRAFTER includes cryo-EM density maps and secondary structure information that are generated from mutate-and-map read out by next generation sequencing (M2-seq). The three techniques M2-seq, cryo-EM, and auto-DRRAFTER comprise the Ribosolve pipeline. All-atom models are generated in auto-DRRAFTER with the accuracy predicted by overall modeling convergence. Uncertain regions are identified by comparing the per-residue-convergence and real-space cross correlation (CC) between the map and the model (Fig. 1). Using auto-DRRAFTER we determined 11 previously unknown protein-free RNA structures ranging from 119 to 388 nucleotides (nt), demonstrating its ability to rapidly and accurately model protein-free RNA structures [11].

2 Materials

2.1 Recommended Computing Requirement

Typical use of auto-DRRAFTER requires a high-performance computing cluster. While runs may be executed interactively, most auto-DRRAFTER runs benefit from using a job scheduler such as SLURM or Torque.

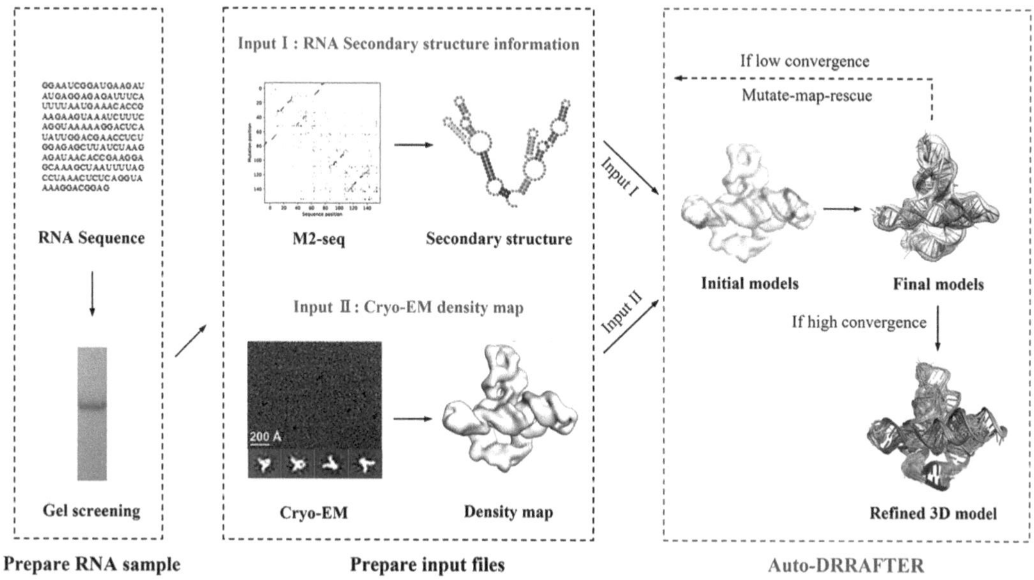

Fig. 1 Auto-DRRAFTER requires RNA secondary structure and cryo-EM density map as input to generate all-atom RNA models

All jobs were executed on SKLB Duyu high performance computing center at Sichuan University and Sherlock 2.0 high performance computing cluster at Stanford University. One RNA model normally takes 2–5 auto-DRRAFTER rounds to converge to the final model building task that takes another two rounds. Each auto-DRRAFTER round contains 150 jobs per helix placement of one end node (described below), with each job using one processor core available on either Duyu or Sherlock (Intel 6230, Intel E5-2640v4, Intel 5118 or AMD 7502). In the case of *F. nucleatum* glycine riboswitch, it took 144 computing hours (a total of six rounds with 24 h per round) on 600 cores (a total of four end nodes with 150 jobs per end node).

2.2 Installing Software

2.2.1 Rosetta (auto-DRRAFTER)

1. The auto-DRRAFTER software is freely available to academic users as part of the Rosetta software suite at www.rosettacommons.org. (*see* **Note 1**).
2. Documentation is available at https://www.rosettacommons.org/docs/latest/application_documentation/rna/auto-drrafter.
3. Demo is available at https://www.rosettacommons.org/demos/latest/public/auto-drrafter/README.
4. A limited version of the software is also freely available through an online ROSIE server at https://rosie.rosettacommons.org/auto-drrafter.

5. Install Rosetta RNA tools. See instructions and documentation here: https://new.rosettacommons.org/docs/latest/application_documentation/rna/RNA-tools.

2.2.2 Python

Make sure that you have Python version 2.7.16 installed, then install networkx (https://networkx.org) and mrcfile (https://mrcfile.readthedocs.io). To check whether the installation is successful, run the "*pip install networkx mrcfile*"command.

2.2.3 UCSF Chimera

Download and install Chimera version 1.1.4 at http://www.cgl.ucsf.edu/chimera/download.html.

2.2.4 EMAN2

EMAN2 is used by auto-DRRAFTER under the hood to process cryo-EM maps. Download and install EMAN2 version 2.31 (Version 2.2 is also available) at https://blake.bcm.edu/emanwiki/EMAN2/Install. And run the "*e2proc3d.py -h*" and "*e2Segment3d.py -h*" commands to check whether e2proc3d.py and e2Segment3d.py are successfully installed.

2.3 Prepare Input Files

The RNA 3D model is derived from the sequence (fasta.txt) and secondary structure in dot-bracket format (secstruct.txt), with cryo-EM density (map.mrc) as additional 3D experimental restraints (*see* Table 1 and files deposited at https://github.com/auto-DRRAFTER/springer-chapter). The RNA secondary structure can be obtained from literature analyses on the target molecule that take into account covariance or, if such data are unavailable, prediction methods such as CONTRAfold (http://contra.stanford.edu/contrafold/), RNAFold (available at http://rna.tbi.univie.ac.at/cgi-bin/RNAWebSuite/RNAfold.cgi), or RNAstructure (available at https://rna.urmc.rochester.edu/RNAstructure.html). More accurate predictions can be achieved through experimental restraints from chemical probing data using selective hydroxyl acylation analyzed by primer extension (SHAPE) [26] or dimethyl sulfate (DMS) [27] mapping, which can be used as inputs in secondary structure prediction algorithms. In the Ribosolve pipeline, RNA secondary structure files are generated from M2-seq, a method that monitors changes in chemical reactivity at each residue as every other residue is mutated. Compared to conventional chemical probing approaches, M2-seq provides more accurate secondary structures as well as uncertainties from nonparametric bootstrapping [28, 29]. We note that high certainty secondary structures are currently a prerequisite to achieving confident tertiary structures from cryo-EM with resolutions worse than 4 Å. Cryo-EM maps are obtained with or without Volta phase plate on a 200 kV or 300 kV microscope. Here we describe a detailed protocol of the Rosetta-based auto-DRRAFTER that utilizes the above mentioned input files to automatically generate all-atom RNA models.

Table 1
A brief explanation of the input files

Input files	Brief explanation
fasta.txt (This example uses FNG_riboswitch.fasta)	The FASTA file, which lists the full sequence of your RNA molecule. It should contain one line that starts with ">" and lists the chain and residue numbers for the sequence. The RNA sequence should be specified with lower-case letters. Currently auto-DRRAFTER can only handle single chain RNAs.
secstruct.txt (This example uses FNG_riboswitch.txt)	A file containing the secondary structure of the complex in dot-bracket notation. The secondary structure should be the same length as the sequence found in the fasta file. This secondary structure will be enforced during the auto-DRRAFTER modeling. We recommend verifying predicted secondary structures biochemically with M2-seq.
map.mrc (This example uses FNG_riboswitch.mrc)	The density map file in MRC format. It can be obtained on 200 kV or 300 kV microscopes by collecting cryo-EM images with or without the use of a Volta phase plate.
job_submission_template.sh	This job submission script uses the conventional directives that specify the resources used to run each round of auto-DRRAFTER and should reflect the job scheduler used on the cluster you are planning to run your job on (*e.g.*, SLURM or TORQUE). If running interactively, the *job_submission_template.sh* file can be an empty file.

3 Methods

3.1 Auto-DRRAFTER Overview

The auto-DRRAFTER program starts by importing the secondary structure information of RNA (Fig. 2a) and the cryo-EM density map (Fig. 2b). Firstly, auto-DRRAFTER identifies end nodes in the density map – regions in which the map contains terminal hairpin loops. To identify these nodes, the density map is low-pass filtered to 20 Å and manually adjusted to the appropriate threshold so that the double-helix regions look like rods with their ends easier to identify. Stem loop regions in the secondary structure are automatically selected as candidates that will be placed at each end node (Fig. 2c), which are labeled as red spheres in the low-pass filtered cryo-EM map (Fig. 2d). Next, a single end node is randomly selected (circled in Fig. 2d) as a starting point to place each stem loop candidate from the secondary structure (black circle regions in Fig. 2c). An idealized A-form helix is generated for each stem loop candidate probe helix (probe helix colored black in Fig. 2e) and automatically placed in the selected end node of the original cryo-EM map (boxed in Fig. 2e). The resulting probe helices in the original cryo-EM maps are used as initial models in the subsequent fragment-based RNA folding method in auto-DRRAFTER to generate all-atom RNA models (Fig. 2f). Hundreds of models are generated in the iterative RNA fragment assembly and refinement

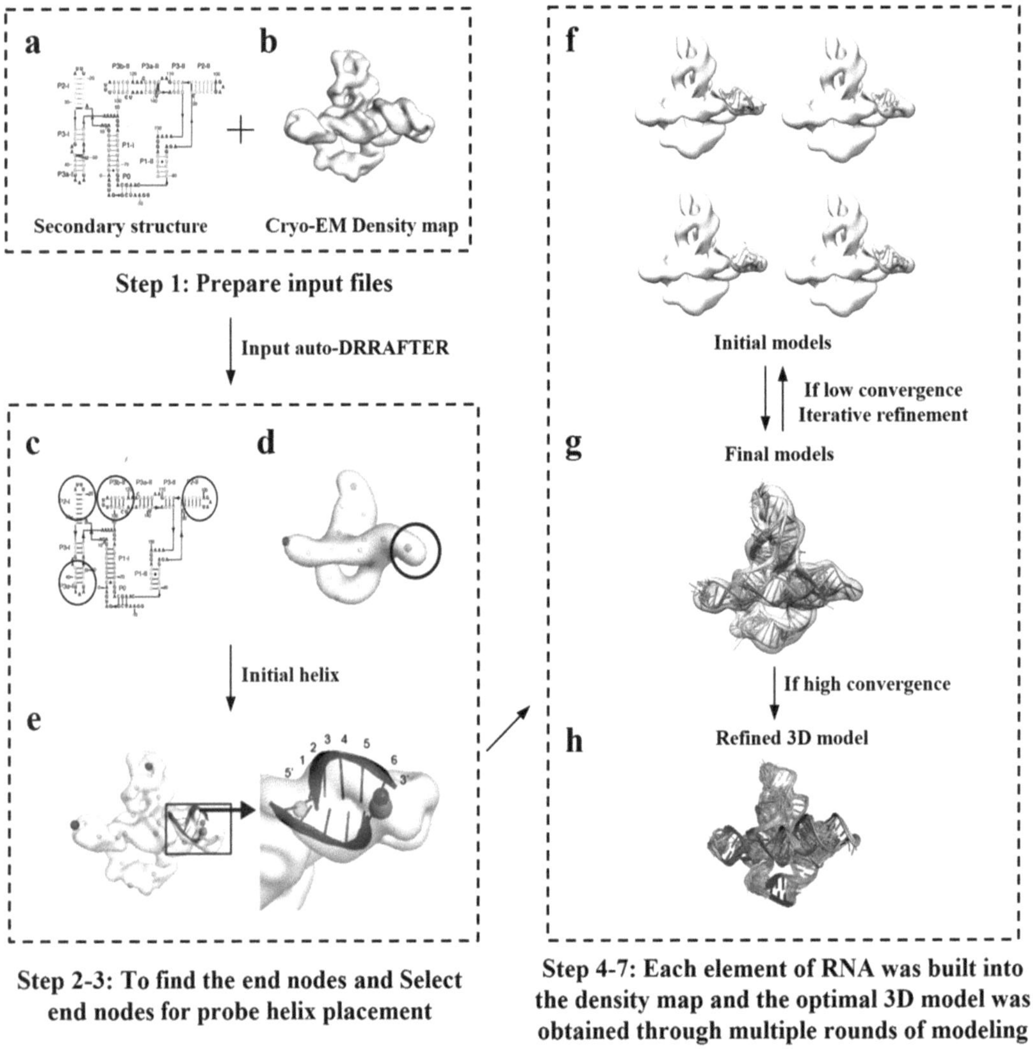

Fig. 2 Auto-DRRAFTER workflow. *F. nucleatum* glycine riboswitch is shown here as an example. RNA secondary structure (**a**) and the cryo-EM map (**b**) are input files for auto-DRRAFTER. Next, end node candidates in the secondary structure (circled in **c**) and end nodes in the low-pass filtered cryo-EM map (red spheres in (**d**) are identified; blue spheres are used to label stems. Then, the end node candidate stem model (probe helix) is placed in the randomly selected end node in the original cryo-EM map and optimized by detecting the distance between the C1′ atom of nucleotide 6 (black sphere) and the end node (black box and black helix in (**e**). All four probe helices serve as starting point to iteratively generate all-atom RNA models till convergence (**f**, **g**). The model with highest overall score yields the final model (**h**)

rounds, and these models are sorted according to a score function that includes both physics-based score terms and the fit between model fragments and the cryo-EM map. Areas with sufficient structural consensus are fixed in the next iteration. This automated process continues until the entire model is confidently built (Fig. 2f, g). Finally, the top models obtained from the last round of modeling is optimized to yield the final model set (Fig. 2h).

3.2 Step-by-Step Method Details

3.2.1 Prepare Input Files Including RNA Secondary Structure (Fig. 2a) and Cryo-EM Density Map (Fig. 2b)

1. The dot-bracket RNA secondary structure file (FNG_riboswitch.txt), which can be generated from prediction software and websites, or from M2-seq protocol that has been previously described [28, 29].
2. Cryo-EM RNA density map (FNG_riboswitch.mrc). Detailed protocol of cryo-EM data processing is described in another chapter.

3.2.2 Low-Pass Filter the Cryo-EM Density Map and Determine an Appropriate Threshold Level (Fig. 2d) to Find End Nodes (Fig. 2c, Black Circles)

1. Go to the project directory and type the command shown below (*see* Table 2):

```
python $ROSETTA/main/source/src/apps/public/DRRAFTER/auto-DRRAFTER_setup.py -map_thr 20 -full_dens_map input_files/FNG_riboswitch.mrc -full_dens_map_res 10.0 -fasta input_files/FNG_riboswitch.fasta -secstruct input_files/FNG_riboswitch.txt -out_pref FNG_riboswitch -rosetta_directory $ROSETTA/main/source/bin/ -nstruct_per_job 200 -cycles 30000 -fit_only_one_helix -rosetta_extension .static.linuxgccrelease -just_low_pass
```

Table 2
A brief explanation of flags on the typed command

Command flags	Brief explanation
-map_thr	The density threshold at which the detection of optimal helix placement locations will take place. This value is a placeholder for now until we find the best value to use in a later step
-full_dens_map	The path to our density map.
-full_dens_map_res	The resolution of the density map in Å.
-fasta	The path to our fasta file listing the sequence of our RNA molecule.
-secstruct	The path to our secondary structure of the RNA molecule in dot-bracket notation.
-out_pref	The arbitrary prefix for output files generated by auto-DRRAFTER.
-rosetta_directory	The location of the Rosetta executables.
-rosetta_extension	The extension of your Rosetta executables. This extension depends on how Rosetta was installed (look in your Rosetta directory to confirm the extension for your executables).
-repeats	The number of independent attempts to place the helices. 10 is usually a good number for this setting.
-shift_center	This is a setting that is required if you are using a simulated density map that was generated with EMAN2. It should not be necessary in other situations, such as our example in this chapter.

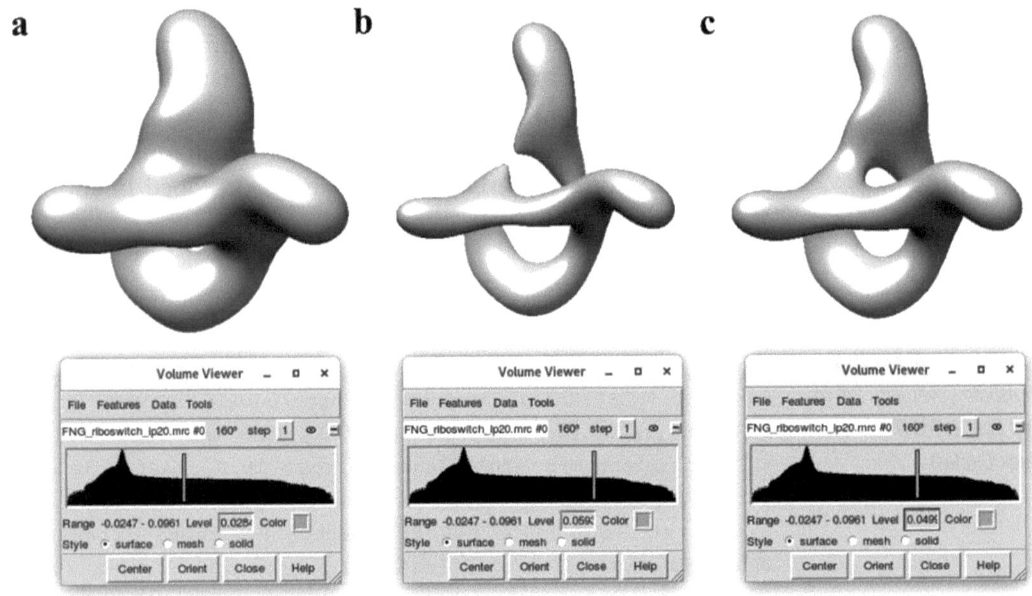

Fig. 3 Use the Volume Viewer tool in Chimera to determine the appropriate threshold level. (**a**) The density threshold shown is lower (0.0284). Although we can see that the end of the spiral may be on this map, we may go to a higher threshold and these areas will be easier to identify. (**b**) The density threshold (0.0593) is too high, and the whole map is no longer connected. (**c**) The threshold (0.0499), works well since the map is still fully connected, and we can clearly see where the ends of the helix/hairpin will be located. Therefore, we will use 0.0499 as the threshold value

Time: Running this command takes less than 20 s (on Intel Gold 6230 2.10 GHz).

2. This will create a single file: FNG_riboswitch_lp20.mrc, indicating that the density map is low-pass filtered to 20 Å, which will be used to figure out the end nodes for candidate placements of probe helices.
3. The appropriate threshold level can be determined using the Volume Viewer Tool in Chimera (Fig. 3). Open the FNG_Riboswitch_lp20.mrc file in Chimera and choose Tools > Volume Data > Volume Viewer. Find the "end node" in the density map that can be clearly identified by changing the threshold value of the map; generally, this value corresponds with the highest threshold value that does not break the continuity of the map. Record the threshold value.
4. This threshold value is only used for placing the probe helix and not for subsequent steps. Here, we used 0.0499 as the threshold value.

3.2.3 Select End Nodes for Probe Helix Placement (Fig. 2d, e)

1. Set up the auto-DRRAFTER run by typing the following command (*see* **Note 2**):

```
python $ROSETTA/main/source/src/apps/public/DRRAFTER/auto-DRRAFTER_setup.py -map_thr 0.0499 -full_dens_map input_files/FNG_riboswitch.mrc -full_dens_map_reso 10.0 -fasta input_files/FNG_riboswitch.fasta -secstruct input_files/FNG_riboswitch.txt -out_pref FNG_riboswitch -rosetta_directory $ROSETTA/main/source/bin/ -nstruct_per_job 200 -cycles 30000 -fit_only_one_helix -rosetta_extension .static.linuxgccrelease
```

Time: about 3–4 min (on Intel Gold 6230 2.10 GHz).

2. After the above command is executed, the information in Fig. 4 will be entered on the screen and a series of files (*see* Table 3) will be generated.
3. This command will generate an *FNG_riboswitch_init_points.pdb* file (*see* Table 3) that consists of spheres fitted throughout the cryo-EM map (Fig. 2d); each sphere has a numeric label. The user can manually open the *FNG_riboswitch_init_points.pdb* file in Chimera or PyMOL and locate the spheres as selected end nodes. For illustration purpose, we colored the potential end node sphere in red and spheres in the middle of

```
Using only one end node!
Low-pass filtering the map to 20A.
Converting density map to graph.
Possible end nodes in the map:
1 5 11
You can visualize the end nodes in FNG_riboswitch_init_points.pdb
You can specify which of these end nodes you'd like to use with -use_end_node
Converting secondary structure to graph.
Mapping secondary structure to density map.
Setting up DRRAFTER runs.
Making full helix H7
Making full helix H4
Making full helix H10
Making full helix H15
```

Fig. 4 The printed information after the command is executed. Firstly, the input density is low-pass filtered to 20 Å and points are placed throughout the density map. These points are then converted into a graph to identify possible end nodes. These end nodes are printed out to the console, which here are nodes 1, 5, and 11. All the points that were placed into the density map are written to a PDB file (here *FNG_riboswitch_init_points.pdb*) that we can visualize in Chimera or PyMOL. Each point has a different number, which corresponds to a node number. Loading the PDB with these points into Chimera (or any other molecular visualization software), residue 1, 5, and 11 can be inspected to see which nodes auto-DRRAFTER has identified as potential end nodes

Table 3
Output files after the command is executed

Output files	Brief explanation
settings_FNG_riboswitch.txt	This file lists all of the settings that were used to set up this auto-DRRAFTER run.
fasta_FNG_riboswitch.txt	The fasta file for the DRRAFTER runs. This is basically the same as the input fasta file, but the numbering has been changed to start at 0 – this is just an auto-DRRAFTER convention. A final step at the end of the auto-DRRAFTER modeling will map these residues numbers back to the input residue numbers.
all_aligned_FNG_riboswitch_0.REORDER.pdb all_aligned_FNG_riboswitch_1.REORDER.pdb all_aligned_FNG_riboswitch_2.REORDER.pdb all_aligned_FNG_riboswitch_3.REORDER.pdb	These are four possible hairpin placements in the density map (visualized above). Each model will be used in a separate DRRAFTER run.
command_FNG_riboswitch_*_R1	The commands for the two DRRAFTER runs for the four possible alignments of the helices into the density map.
flags_FNG_riboswitch_*_R1	These files contain all the flags that will be used for the four DRRAFTER runs for the different helix alignments in the density map.
FNG_riboswitch_H*.pdb	Ideal A-form helices for helix.
FNG_riboswitch_H*_full.out.1.pdb	The same ideal A-form helices from above with placeholder coordinates added for the hairpins.
FNG_riboswitch_auto_fits.txt	This file lists all of the different possible alignments of helices into the density map. Each alignment is numbered and listed on a separate line in this file.
FNG_riboswitch_init_points.pdb	This PDB file contains all of the points that were placed into the density map in order to convert the density map into a graph. This is the file that we visualized later in the chapter to check the assignment of end nodes.
secstruct_FNG_riboswitch.txt	The secondary structure file for the DRRAFTER runs.

stems in blue (Fig. 5). Auto-DRRAFTER randomly chooses one of these end nodes (red spheres) to place the probe helices; however, if a specific end node is preferred, the user may append the flag "*-use_end_node*" at the end of above command.

4. This command also converts the RNA secondary structure (that we specified) to a graph in which helices are represented as edges and junctions and loops are represented as nodes. This secondary structure graph is then mapped onto the graph for the density map. The location of the probe helix was optimized

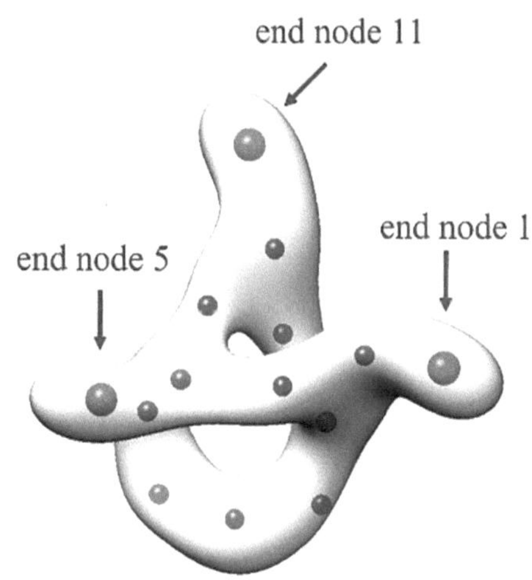

Fig. 5 Visualize all spheres placed in the density map in Chimera. Each point has a numerical label, which corresponds to a node number. We enlarged and colored the end nodes red and other spheres in the middle of stems blue

by minimizing the distance between the C1′ atom of 6 nucleotide in the probe helix and the end node sphere (Fig. 2e). Four possible hairpin placements can be checked in Chimera by opening the density map along with the helix coordinates which are in PDB files with the prefix "*all_aligned_FNG_riboswitch*" (Fig. 6). It is essential to ensure that the helices are placed well in the density map. If the fitting is not appropriate, the remaining steps will likely not yield reasonable results.

3.2.4 Build Initial Models with Different End Nodes (Fig. 2f)

1. Auto-DRRAFTER then continues to build a complete RNA model by stochastically placing the remaining helices and nucleotides from the secondary structure into the cryo-EM density map.
2. Type and run the following command (*see* Table 4):

```
python $ROSETTA/main/source/src/apps/public/DRRAFTER/submit_jobs.py -out_pref FNG_riboswitch -curr_round Ri -njobs 150 -template_submission_script input_files/job_submission_template.sh -queue_command sbatch
```

Time: Typically, one round of modeling takes 12–48 h, depending on the number of models built for each job and how much computing resources are used (*see* Subheading 2.1, Recommended Computing Requirement).

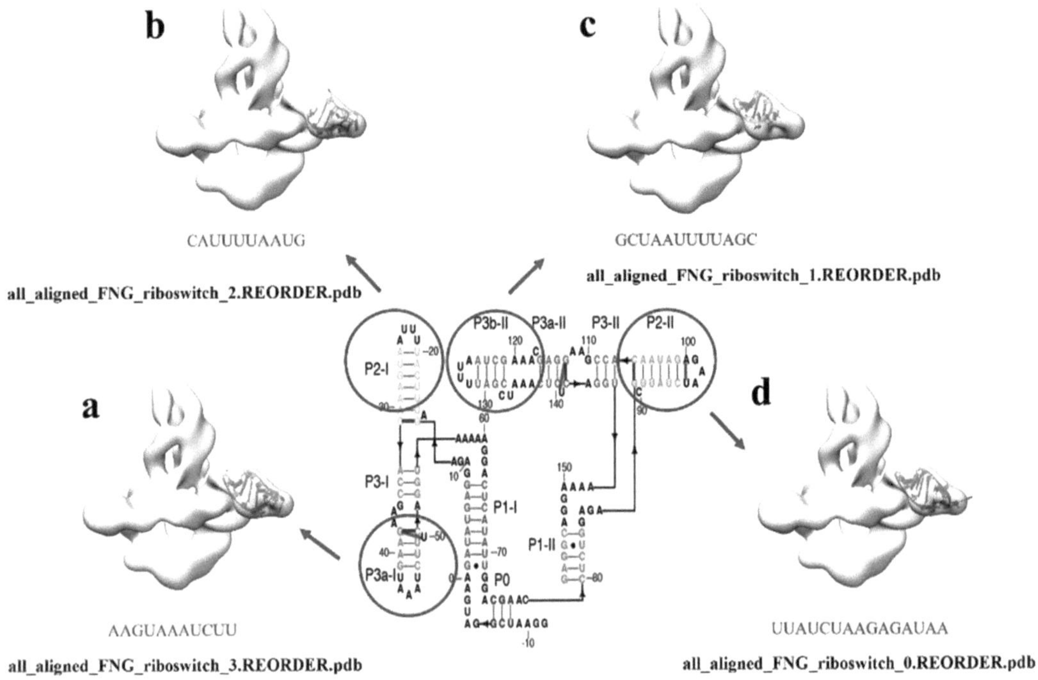

Fig. 6 Visualize the density map of helices in Chimera to check that helices are properly positioned. Four hairpins in the secondary structure are mapped to the end node (we use end node 1 as an example in the figure.). All candidate hairpin placements in the end node will be used to generate models, and an optimized placement will be selected later. (**a**) The corresponding sequence is: AAGUAAAUCUU. (**b**) The corresponding sequence is: CAUUUUAAUG. (**c**) The corresponding sequence is: GCUAAUUUUAGC. (**d**) The corresponding sequence is: UUAUCUAAGAGAUAA

3. This will create the following output directories and files:

job_files: This is a directory that contains all the job submission files.

out_FNG_riboswitch_n_Ri: This directory contains all the DRRAFTER models built starting from probe helix n, where n is an integer to indicate the probe helix number; i indicates the round number.

3.2.5 Iterative Refinement of RNA Models (Fig. 2f, g)

1. Collect all models built in the previous step and calculate overall convergence using the command shown below (*see* **Note 3**):

```
python $ROSETTA/main/source/src/apps/public/DRRAFTER/auto-DRRAFTER_setup_next_round.py -out_pref FNG_riboswitch -curr_round Ri -rosetta_directory $ROSETTA/main/source/bin/ -convergence_threshold 10 -rosetta_extension .static.linuxgccrelease
```

Table 4
A brief explanation of flags in the command above

Flags	Brief explanation
-out_pref	The prefix used for all output files from this run. This should be the same –out_pref that was used in the setup command in Subheadings 3.2.2 and 3.2.3.
-curr_round	This is the round of modeling that is currently being performed. We have not done any modeling yet, so this is round 1. This should always be "R" followed by a number that indicates the round number.
-njobs	This is the number of jobs that will be run. The number of jobs submitted by "-njobs" depends on the real modeling scenario, for a very simple system we used two jobs, but for larger systems we typically use more than 100 jobs. Each job will build the number of models that was specified in the setup command above (–nstruct_per_job), in this case 200 structures per job. We will therefore build 30,000 models in total (200 structures per job × 150 jobs).
-template_submission_script	This flag points to the path of the job submission script, which specifies the resources that auto-DRRAFTER jobs will have access to.
-queue_command	This is the command that will be used to submit the job files to a cluster queuing system. The value for "-queue_command" flag depends on the job scheduler used. For TORQUE computing clusters, the value for the flag would be "qsub" while for SLURM-based clusters, "sbatch" would be used instead. Conversely, "source" is used to run a command interactively. In this example, we use "sbatch" since our computing cluster is based on SLURM.

2. After executing the above command, "Overall Convergence" and "Density Threshold" (*see* **Note 3** and Fig. 7a) are displayed on the screen as well as output files generated (*see* Table 5). This step collects all of the models from the previous step and calculates the convergence of the overall top ten scoring models (across all alignments). Then the next round of modeling is set up based on the models that were built from the previous round. Regions of the models that are well converged will be kept fixed in the next round of modeling.
3. This step outputs several files (*see* Table 5), of which the "FNG_riboswitch_all_models_all_fits_Ri.out.*" PDB files are the overall best scoring models from all of the different alignments in the first round of modeling. We should pay attention to the overall convergence score and how these models fit in the density map (Fig. 7a, b). The first round models do not need to look perfect, but they should roughly fit in the density map.
4. If the overall convergence value is larger than 10 Å, multiple iterations (Subheadings 3.2.4 and 3.2.5) need to be performed

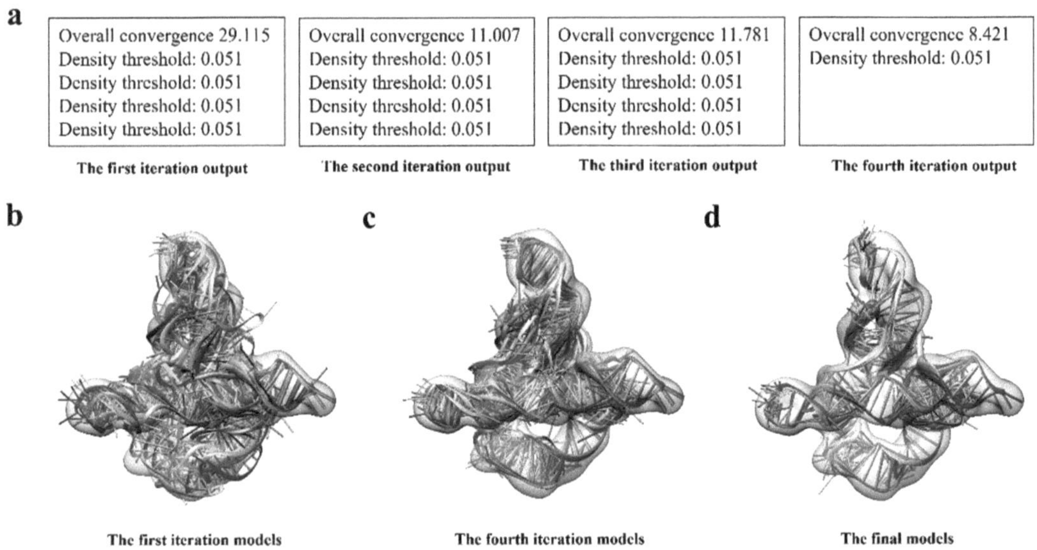

Fig. 7 The main output results in the six rounds of modeling. (**a**) The Overall Convergence and Density Threshold values printed after each iteration. (**b**) Visualize the 10 models generated in the first round of modeling in Chimera to check if models are properly positioned. (**c**) The 10 models generated in round 4 modeling. (**d**) The final models generated in round 6 modeling

Table 5
The output file after the command is executed

Output files	Brief explanation
FNG_riboswitch_all_models_all_fits_Ri.out.*.pdb	These are the overall best scoring models from this round of modeling. It is recommended that we visualize these models in the density map after each round.
command_FNG_riboswitch_Ri	This is the command file for the next round of modeling
flags_FNG_riboswitch_Ri	This file contains all the flags that will be used for each DRRAFTER run for the different helix alignments in the density map.
all_fit_FNG_riboswitch_Ri.REORDER.pdb	This PDB file contains initial helix placements for the next round of modeling.
convergence_FNG_riboswitch_all_models_Ri.txt	This file lists the convergence for the overall top scoring models.
FNG_riboswitch_Ri_CAT_ALL_ROUNDS.out.*.pdb	These are the top scoring models from each of the possible initial helix placements.

until the convergence value is less than 10 Å. In each of these modeling iterations, regions of the RNA models that have converged are kept fixed to focus sampling on segments that remain highly variable.

5. In the FNG_riboswitch example, four iterations are needed by modifying i = 1,2,3 and 4. After four iterations, the overall convergence value was 8.421, and the Density threshold value was 0.051 (Fig. 7a). Visualize the FNG_riboswitch_all_models_all_fits_R4.out.*.pdb file in Chimera to check the 10 models generated (Fig. 7c). Please note that all numbers here (*e.g.*, convergence score, number of iterations) may vary in different trials of the exact same modeling task.

3.2.6 Final Two Rounds of Modeling (Fig. 2g)

1. When the overall convergence of models is less than 10 Å, the user can proceed to the final two rounds of modeling (*see* **Note 4**).
2. In the first final round, all regions of the RNA are allowed to move from their initial positions to achieve further refinement of the modeled coordinates. Run the command below to begin the first final round of modeling (*see* **Note 5**):

```
python $ROSETTA/main/source/src/apps/public/DRRAFTER/submit_jobs.py -out_pref FNG_riboswitch -curr_round FINAL_R5 -njobs 150 -template_submission_script input_files/job_submission_template.sh -queue_command sbatch
```

3. Set up the last round of modeling with the following command. This command extracts converged regions from the models with highest scores in the previous iteration, setting up the last round of modeling (*see* **Note 6**):

```
python $ROSETTA/main/source/src/apps/public/DRRAFTER/auto-DRRAFTER_setup_next_round.py -out_pref FNG_riboswitch -curr_round FINAL_R5 -rosetta_directory $ROSETTA/main/source/bin/ -rosetta_extension .static.linuxgccrelease
```

4. In the second final round, converged regions from the previous round are held fixed during fragment assembly steps of auto-DRRAFTER, and they are only allowed to move during high-resolution refinement steps. Run the final round of modeling with the following command:

```
python $ROSETTA/main/source/src/apps/public/DRRAFTER/submit_jobs.py -out_pref FNG_riboswitch -curr_round FINAL_R6 -njobs 150 -template_submission_script input_files/job_submission_template.sh -queue_command sbatch
```

5. Retrieve the top-scoring final models using the command below:

```
python $ROSETTA/main/source/src/apps/public/DRRAFTER/auto-
DRRAFTER_setup_next_round.py -out_pref FNG_riboswitch -curr_-
round FINAL_R6 -rosetta_directory $ROSETTA/main/source/bin/
-rosetta_extension .static.linuxgccrelease
```

6. The output of this command is: DONE building models for FNG_riboswitch.

3.2.7 Completing the Modeling Procedure

1. Type the command shown below (*see* **Note 7**):

```
python $ROSETTA/main/source/src/apps/public/DRRAFTER/finali-
ze_models.py -fasta input_files/FNG_riboswitch.fasta -out_-
pref FNG_riboswitch -final_round FINAL_R6
```

2. This should print "Done finalizing models" to the screen, indicating that the modeling is complete. This creates final models (Fig. 7d): *FNG_riboswitch_all_models_all_fits_FINAL_R6.out.*.pdb*.
3. The final models should be carefully examined and validated. If there are large parts of the model that are outside the density map, or large regions of the map have missing coordinates, the model is likely not properly built. In this case, some additional attempts may be useful, such as specifying a different end node as the initial helical probe for remodeling, verifying the secondary structure of RNA through further experiments, or modeling by using the more definitive structure as an initial position and so on. In addition, the obtained model can be verified in a variety of ways, with additional biochemical or functional experiments, or through comparison to homologous structures or any available crystal structures of subregions.
4. We can use the linear relationship between modeling convergence and predicted root mean square deviation (RMSD) from the native structure (estimated accuracy = 0.61× convergence +2.4) to estimate model accuracy using convergence value. The auto-DRRAFTER modeling convergence, defined as the average pairwise RMSD of the top ten scoring models, should be correlated with the model RMSD. In this case, the overall convergence of the final ten best models after the last round (FINAL_R6) of modeling is 2.833 Å, yielding the final estimated accuracy of 4.128 Å.

4 Notes

1. Rosetta is free for academic users, but you need permission before you can download it (https://els2.comotion.uw.edu/product/rosetta). auto-DRRAFTER is available in the Rosetta

weekly releases after 2019.47 (it is not available in 2019.47) and files for this example are available in releases after 2020.03. auto-DRRAFTER is not available in Rosetta 3.11.

2. This is the same command we used in the previous step, with the -just_low_pass flag removed and the "-map_thr" updated with the value we recorded in Subheading 3.2.2. Depending on where the helix appears, we may need to try different thresholds. In this particular case, "-map_thr 0.0555" also yielded a similar result.
3. The default and recommended value for "convergence_threshold" is 10. To try out the Ribosolve workflow for the first time, a more lenient threshold of 100 may be used to reduce runtime and assess whether the installation is successful. The "Density threshold" is a critical value that is determined automatically, which is used to determine which regions of the RNA are sufficiently converged to be kept fixed during the next round of modeling. The value that is printed to the screen is the value that was determined automatically. Sometimes, particularly for higher-resolution maps (better than about ~6 Å resolution) or for maps that are relatively noisy, this automatically selected value is too high. This can make it so that regions that are well converged and do fit reasonably well in the density map are not kept fixed in subsequent rounds of modeling, making it much more challenging for the models to converge overall. You can check whether this might be a problem in your case by examining the top scoring models from each fit from the last round (FNG_riboswitch_Ri_CAT_ALL_ROUNDS.out.*.pdb) as well as the PDB files containing the regions that will be kept fixed for the subsequent round of modeling (all_fit_FNG_riboswitch_Ri.REORDER.pdb). If there are regions of your models that look sufficiently converged, but do not show up in the "all_fit_FNG_riboswitch_Ri.REORDER.pdb" files, then you might want to rerun this step and use the flag "-dens_thr" to select a lower value for the density threshold. It often makes sense to try reducing the density threshold by ~50%. If you still see a similar problem, you can reduce the density threshold even further.
4. When the convergence has reached below 10 Å, the auto-DRRAFTER_setup_next_round.py command will produce a command file with the FINAL keyword, indicating that the user can proceed to the final two rounds of modeling. If the convergence remains over 10 Å and no longer improves, the auto-DRRAFTER program will skip to the final modeling.
5. This is basically the same command from Subheading 3.2.4 except "–curr_round" has changed to FINAL_R5. Note that if the modeling has not reached the FINAL stages, "R5" should

be used in place of "FINAL_R5." This creates files very similar to those from Subheading 3.2.5. Again, we should check the models that we built (FNG_riboswitch_all_models_all_fits_FINAL_R5.pdb).

6. In this example, the output of this command is: Overall convergence 4.484, density threshold: 0.051. Convergence was 4.0 in the original Ribosolve pipeline [11].
7. -final_round is the round of modeling that was just completed, in this case FINAL_R6.

Acknowledgments

Computational tasks were performed on SKLB Duyu high performance computing center in Sichuan University and Sherlock 2.0 high performance computing cluster in Stanford University. This work was supported by Ministry of Science and Technology of China (MoST) 2021YFA1301900, Natural Science Foundation of China (NSFC) 82041016 and 32070049, and Sichuan University start-up funding 20822041D4057 to Z.S., the National Science Foundation (NSF) Graduate Research Fellowship award DGE-1656518 to R.R., the Gerald J. Lieberman Fellowship to R. R., the Rosetta REU NSF award 1950697 to P.P., and the National Institutes of Health R35 grant GM122579 to P.P. and R.D.

References

1. Atkins JF, Gesteland R, Cech T (2011) RNA worlds: from life's origins to diversity in gene regulation. Cold Spring Harbor Laboratory Press, New York
2. Cech Thomas R, Steitz, Joan A (2014) The noncoding RNA revolution – trashing old rules to forge new ones. Cell 157:77–94
3. Blundell TL, Chaplin AK (2021) The resolution revolution in X-ray diffraction, Cryo-EM and other technologies. Prog Biophys Mol Biol 160:2–4
4. Watson ZL, Ward FR, Meheust R et al (2020) Structure of the bacterial ribosome at 2 A resolution. elife 9:e60482
5. Su Z, Zhang K, Kappel K et al (2021) Cryo-EM structures of full-length Tetrahymena ribozyme at 3.1 A resolution. Nature 96(7873):603–607
6. Schlick T, Pyle AM (2017) Opportunities and challenges in RNA structural modeling and design. Biophys J 113:225–234
7. Sripakdeevong P, Beauchamp K, Das R (2012) Why can't we predict RNA structure at atomic resolution? In: Leontis N, Westhof E (eds) RNA 3D structure analysis and prediction. Springer, Berlin/Heidelberg
8. Zhang J, Ferre-D'amare AR (2014) New molecular engineering approaches for crystallographic studies of large RNAs. Curr Opin Struct Biol 26:9–15
9. Zhang H, Keane SC (2019) Advances that facilitate the study of large RNA structure and dynamics by nuclear magnetic resonance spectroscopy. Wiley Interdiscip Rev RNA 10:e1541
10. Hura GL, Menon AL, Hammel M et al (2009) Robust, high-throughput solution structural analyses by small angle X-ray scattering (SAXS). Nat Methods 6:606–612
11. Kappel K, Zhang K, Su Z et al (2020) Accelerated cryo-EM-guided determination of three-dimensional RNA-only structures. Nat Methods 17:699–707
12. Jossinet F, Ludwig TE, Westhof E (2010) Assemble: an interactive graphical tool to analyze and build RNA architectures at the 2D and 3D levels. Bioinformatics 26:2057–2059

13. Parisien M, Major F (2008) The MC-fold and MC-Sym pipeline infers RNA structure from sequence data. Nature 452:51–55
14. Xia Z, Bell DR, Shi Y et al (2013) RNA 3D structure prediction by using a coarse-grained model and experimental data. J Phys Chem B 117:3135–3144
15. Liebschner D, Afonine PV, Baker ML et al (2019) Macromolecular structure determination using X-rays, neutrons and electrons: recent developments in Phenix. Acta Crystallogr D Struct Biol 75:861–877
16. Das R, Baker D (2007) Automated *de novo* prediction of native-like RNA tertiary structures. Proc Natl Acad Sci USA 104:14664–14669
17. Xu X, Zhao C, Chen SJ (2019) VfoldLA: a web server for loop assembly-based prediction of putative 3D RNA structures. J Struct Biol 207:235–240
18. Popenda M, Szachniuk M, Antczak M et al (2012) Automated 3D structure composition for large RNAs. Nucleic Acids Res 40:e112
19. Jonikas MA, Radmer RJ, Laederach A et al (2009) Coarse-grained modeling of large RNA molecules with knowledge-based potentials and structural filters. RNA 15:189–199
20. Boniecki MJ, Lach G, Dawson WK et al (2016) SimRNA: a coarse-grained method for RNA folding simulations and 3D structure prediction. Nucleic Acids Res 44:e63
21. Sharma S, Ding F, Dokholyan NV (2008) iFoldRNA: three-dimensional RNA structure prediction and folding. Bioinformatics 24: 1951–1952
22. Miao Z, Adamiak RW, Antczak M et al (2020) RNA-puzzles round IV: 3D structure predictions of four ribozymes and two aptamers. RNA 26:982–995
23. Li B, Cao Y, Westhof E et al (2020) Advances in RNA 3D structure modeling using experimental data. Front Genet 11:574485
24. Townshend RJL, Eismann S, Watkins AM et al (2021) Geometric deep learning of RNA structure. Science 373:1047–1051
25. Kappel K, Liu S, Larsen KP et al (2018) De novo computational RNA modeling into cryo-EM maps of large ribonucleoprotein complexes. Nat Methods 15:947–954
26. Merino EJ, Wilkinson KA, Coughlan JL et al (2005) RNA structure analysis at single nucleotide resolution by selective 2′-hydroxyl acylation and primer extension (SHAPE). J Am Chem Soc 127:4223–4231
27. Peattie DA, Gilbert W (1980) Chemical probes for higher-order structure in RNA. Proc Natl Acad Sci USA 77:4679–4682
28. Cordero P, Kladwang W, Vanlang CC et al (2014) The mutate-and-map protocol for inferring base pairs in structured RNA. Methods Mol Biol 1086:53–77
29. Cheng CY, Kladwang W, Yesselman JD et al (2017) RNA structure inference through chemical mapping after accidental or intentional mutations. Proc Natl Acad Sci USA 114:9876–9881

Chapter 14

X-Ray Crystallography to Study Conformational Changes in a TPP Riboswitch

Ashok Nuthanakanti, Ascensión Ariza-Mateos, and Alexander Serganov

Abstract

Conformational rearrangements are key to the function of riboswitches. These regulatory mRNA regions specifically bind to cellular metabolites using evolutionarily conserved sensing domains and modulate gene expression via adjacent downstream expression platforms, which carry gene expression signals. The regulation is achieved through the ligand-dependent formation of two alternative and mutually exclusive conformations involving the same RNA region. While X-ray crystallography cannot visualize dynamics of such dramatic conformational rearrangements, this method is pivotal to understand RNA–ligand interaction that stabilize the sensing domain and drive folding of the expression platform. X-ray crystallography can reveal local changes in RNA necessary for discriminating cognate and noncognate ligands. This chapter describes preparation of thiamine pyrophosphate riboswitch RNAs and its crystallization with different ligands, resulting in structures with local conformational changes in RNA. These structures can help to derive information on the dynamics of the RNA essential for specific binding to small molecules, with potential for using this information for developing designer riboswitch–ligand systems.

Key words TPP riboswitch, Conformational change, X-ray crystallography, Antibiotic, RNA

1 Introduction

Riboswitches are typically *cis*-acting RNA elements specifically binding cellular metabolites or other small molecules and modulating the expression of genes in all kingdoms of life [1–4]. In bacteria, riboswitches usually reside in the 5′ untranslated regions (UTRs) of messenger RNAs and employ variable mechanisms for regulation of gene expression on the levels of transcription, translation, and mRNA stability [1, 5, 6]. The majority of riboswitches participate in feedback regulation of a riboswitch ligand through inhibition or activation of the downstream genes for enzymes or transporters associated with this ligand. Riboswitches typically contain an evolutionarily conserved aptamer domain, recognizing and binding target metabolites, and a variable expression platform, located downstream of the aptamer domain and responding allosterically

Jienyu Ding et al. (eds.), *RNA Structure and Dynamics*, Methods in Molecular Biology, vol. 2568,
https://doi.org/10.1007/978-1-0716-2687-0_14,

to the metabolite binding [7, 8]. Metabolite-sensing domains adopt a variety of three-dimensional structures tuned for exquisite binding to particular ligands [9], which range from a simple anion F^- [10] to much larger vitamin adenosylcobalamin [11].

The ability of riboswitches to specifically bind to small drug-like molecules raised a question on the potential utility of these controlling elements as targets for novel types of antibiotics [12]. The prevalence of riboswitches in pathogenic bacteria, their association with genes critical for vital cellular processes, and the lack of riboswitches in mammals, make them ideal potential targets for developing of new antibiotics. This idea received further support from observations that several riboswitches can bind to natural antibiotics or synthetic inhibitors mimicking natural riboswitch ligands [13–22]. Recently, Merck developed a synthetic compound, called ribocil, interacting with flavin mononucleotide (FMN) riboswitch and suppressing the growth of pathogenic bacteria [23]. However, enthusiasm for this compound was hampered by the high frequency of resistance mutations in the riboswitch sequence that made the compound impractical in the monotherapy [23, 24].

Despite a drawback with the antibacterial targeting of the FMN riboswitch, other types of riboswitches may still be interesting targets for antibacterials. For example, some bacteria may have several copies of the same type of a riboswitch controlling several genes or gene operons. Simultaneous inhibition of several riboswitch targets may decrease the likelihood of developing resistance. Alternatively, in addition to riboswitches, metabolite-like drugs may also bind to enzymes not controlled by the riboswitch and inhibit their activity. Such simultaneous turning genes off and inhibition of enzymes can lead to a dual antibacterial action.

Regardless of the success of riboswitches as genuine drug targets, riboswitches offer unprecedented views on how RNA binds to small molecules and how the binding of a small molecule allosterically modulates gene expression. Given a recent push for developing RNA-specific drugs to treat human diseases and infections [25, 26], riboswitches represent invaluable test systems for understanding the fundamental principles of RNA folding, dynamics, and intermolecular interactions.

One of the most interesting and widely spread riboswitch types is the thiamine pyrophosphate (TPP)-specific riboswitch [4, 27]. This riboswitch is one of the most abundant classes of riboswitches in bacteria [28] and the only active riboswitch identified in eukaryotic organisms [29–31], such as the ascomycete fungi *Aspergillus nidulans* and *Neurospora crassa* [32–34], the green alga *Chlamydomonas reinhardtii* [35], and the higher plant *Arabidopsis thaliana* [36, 37]. In eukaryotes, TPP riboswitches were found in introns or UTRs and were implicated in alternative splicing that affects translation and stability of mRNAs [32, 34–39]. In bacteria,

TPP riboswitches are located in 5′ UTRs, where they prevent translation by sequestering the ribosome binding site or facilitate transcription termination by promoting formation of the transcription terminator [4, 40]. In all cases, TPP binding to the mRNA downregulates the expression of thiamine-associated genes and thereby decreases TPP biosynthesis.

TPP, the ligand for TPP riboswitches, is an active form of vitamin B_1 and an essential cofactor for many enzymes involved in the amino acid and carbohydrate metabolism. TPP consists of an aminopyrimidine (aPy) ring, a central thiazole (Thz) ring, and a pyrophosphate group (PP) (Fig. 1a). The bacterial TPP riboswitches demonstrate a high affinity (K_D = 8 nM) and strong selectivity towards TPP, discriminating it efficiently from precursors or analogues like thiamine monophosphate (TMP) or pyrithiamine (PT), which bind with 100- and 1000-fold lower affinity, respectively [4, 22].

Several phylogenetic [31, 41], biochemical [4, 31, 42], and structural studies of bacterial and eukaryotic TPP riboswitches [43–46] provided insights on the three-dimensional structure of the riboswitch and explained its high specificity for TPP. A typical TPP riboswitch has 5 helices (P1-P5) and two loops (L3 and L5) connected by joining regions (*e.g.*, J1/4) (Fig. 1b). These elements build two parallel helical domains, the pyrophosphate and the

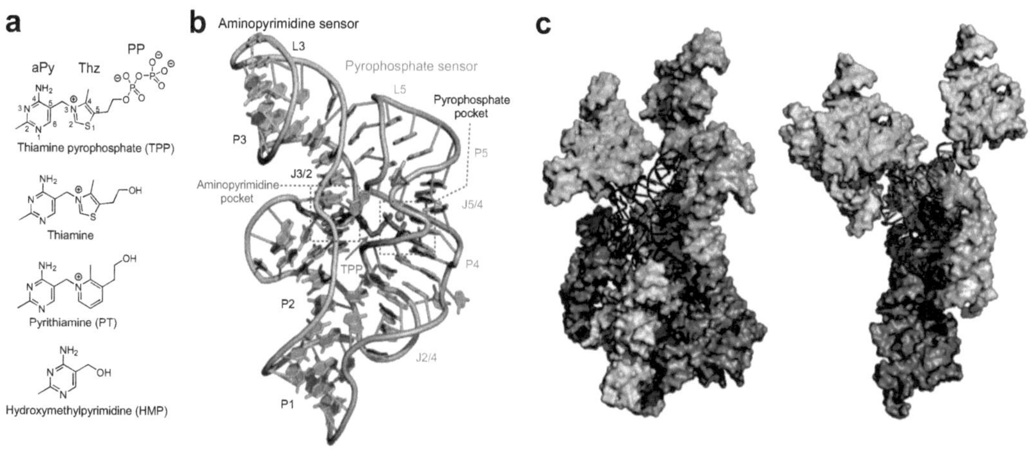

Fig. 1 TPP riboswitch and its ligands. (**a**) TPP and its precursors and analogs. aPy, aminopyrimidine; Thz, thiazole; and PP, pyrophosphate moieties. PT, pyrithiamine; HMP, 4-amino-5-hydroxymethyl-2-methylpyrimidine (HMP). (**b**) X-ray crystal structure of the *E. coli thiM* riboswitch (PDB ID: 2GDI) [44]. The view shows molecule X from the asymmetric unit that contains two molecules. TPP is shown in stick representation with atomic colors (carbon, magenta; oxygen, red; nitrogen, blue; sulfur, yellow; and phosphorus, orange). (**c**) Crystal packing of the TPP-bound TPP riboswitch from Ref. [44] (left) and [45] (right). The central molecule, shown in sticks, is surrounded by symmetry mates at ≤4 Å distance, shown in surface representation. Note that the 2GDI structure contains two molecules in the asymmetric unit (left). The first molecule is shown in sticks and magenta color and the second molecule is shown in surface representation and pink color. The 2HOJ structure contains one RNA molecule in the asymmetric unit (right)

pyrimidine sensors, connected to the switch helix P1 by a three-way junction. TPP binding promotes formation of the P1 helix from the 5′ and 3′ ends of the metabolite-sensing domain and prevents participation of the 3′ sequence in the alternative downstream structure, thereby affecting translation and transcription of the mRNA. Depending on the RNA used for crystallization, the TPP riboswitch can be crystallized in two different forms [44, 45] with distinct crystal packing interactions (Fig. 1c), nevertheless producing very similar overall RNA folds with small differences in the ligand-binding pockets.

In the structure, TPP is perpendicularly positioned to the two helical domains of the riboswitch and uses two pockets located in each helical segment of the riboswitch to grip each end of the compound [44, 45] (Fig. 2a). A pair of Mg^{2+} cations mediate interactions between RNA and the pyrophosphate moiety of TPP. The aminopyrimidine ring is accommodated between the bases of G42 and A43 where it forms hydrogen bonding interactions with adjoining residues G40 and G19. The central thiazole moiety is not involved in extensive recognition, revealing why the antimicrobial compound pyrithiamine pyrophosphate (PTPP) binds to TPP riboswitches.

TPP riboswitches were crystallized in complex with precursors of TPP, like TMP and thiamine, analogues of TPP, such as PT, benfothiamine (BTP), PTPP, and oxythiamine pyrophosphate [45, 48], and a variety of small molecules from a fragment-based screening study [49, 50]. Despite similarities in structures and function, a comparative study revealed differences in the dynamical behavior of the TPP riboswitches in the pyrimidine-binding pocket in bacterial (*E. coli*) and eukaryotic (*A. thaliana*) TPP riboswitches [51], suggesting that this riboswitch class represents an exciting model for elucidating RNA dynamics. Such studies could also benefit from multiple X-ray structures of the riboswitch, which revealed conformational changes associated with binding to different ligands.

In this chapter, we present the methodology to crystallize the TPP riboswitch from the *E. coli thiM* gene with four ligands (Fig. 1a). This riboswitch binds different ligands and can also crystallize in two different forms, thus offering multiple ways to study ligand-dependent conformational changes and RNA dynamics. We describe crystallization of two riboswitch RNAs [44, 45] with TPP to obtain two different crystal forms characterized by a change in the position of G72 in the ligand binding pocket (Fig. 2b). We also describe crystallization of one riboswitch RNA with thiamine, PT, and HMP. All four compounds share the aminopyrimidine moiety, bound to a pyrimidine sensor stem of the TPP aptamer. However, the lack of pyrophosphate moiety in thiamine, PT, and HMP apparently causes conformational changes in G72 and inward movements of nucleotides of the pyrophosphate

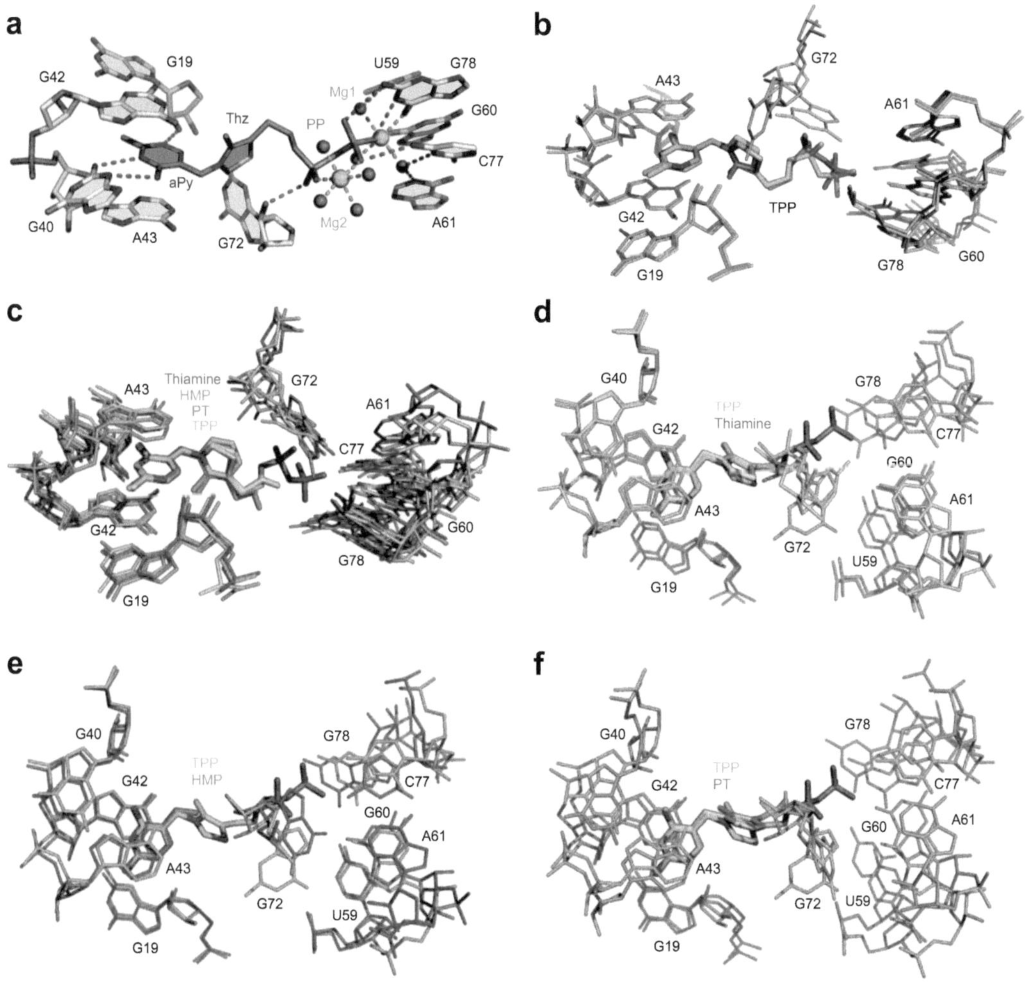

Fig. 2 Details of ligand binding to the *thi*M TPP riboswitch. (**a**) Recognition of TPP. Gray dashed lines depict hydrogen bonds. Mg^{2+} cations and their coordination bonds are shown with green spheres and sticks (PDB ID: 2GDI) [44], respectively. Water molecules are shown with red spheres. (**b**) Superposition of the TPP-bound structures prepared with two different RNAs from Refs. [44, 45]. Green, PDB ID: 2GDI [44]; and violet (PDB ID code: 7TDA [47]). The superposition shows a change in the position of G72. (**c**) Superposition of the ligand-bound TPP riboswitches prepared with the RNA from Ref. [45] and centered on the aminopyrimidine ring. TPP, green and atomic colors; thiamine, yellow; HMP, blue; PT, pink. The view depicts conformational differences in the nucleotides of the pyrophosphate sensor and G72. (**d–f**) Top views of the superposition in panel (**c**) shown for TPP and thiamine (**d**), TPP and HMP (**e**), and TPP and PT (**f**). The images highlight accommodations of G72 and nucleotides of the pyrophosphate sensor (U59, G60, A61, C77, and G78) to bind the ligands of different size

sensor (U59, G60, A61, C77, G78) (Fig. 2c–f). Thus, despite using predetermined crystal packing interactions, X-ray crystallography can offer snapshots on the ligand-dependent conformational rearrangements in RNA and guide single-molecule, molecular dynamics, and other studies to elucidate behavior of RNA molecules in solution. These studies will help us understand the dynamic landscape of RNA and develop new types of RNA binders that can trap metastable RNA conformations, guide RNA folding to different pathways, or induce designer allosteric changes.

2 Materials

2.1 RNA Preparation

2.1.1 DNA Template Preparation

1. 100 mg/mL ampicillin (Amp). Dissolve in 50% ethanol and store at −20 °C.
2. Lysogeny broth (LB)/Amp medium and LB-agar/Amp plates: Dissolve 10 g tryptone, 5 g yeast extract, and 10 g NaCl in 1 L of bi-distillated water. To prepare LB-agar for plates, add 4 g LB-agar to 100 mL of bi-distillated water. Sterilize media by autoclaving for 30 min, and add ampicillin to 100 μg/mL (*see* **Note 1**).
3. PureLink HiPure Expi plasmid Gigaprep kit (Invitrogen, Waltham, MA, USA).
4. Water bath at 37 °C.
5. *Hind*III-HF (100,000 U/mL) and *Bam*HI-HF (20,000 U/mL) restriction enzymes (New England Biolabs, Ipswich, MA, USA) supplied with 10× CutSmart buffer.
6. Ultrapure agarose (Invitrogen, Waltham, MA, USA).
7. Gel chamber, gel casting tray, tooth comb, and other supplies for agarose gels.
8. 50× TAE buffer: 2 M Tris-acetate, pH 8.0, 50 mM ethylenediaminetetraacetic acid (EDTA).
9. 6× Gel Loading Dye for agarose gels, Purple (New England Biolabs, Ipswich, MA, USA).
10. UltraPure Phenol:chloroform:isoamyl alcohol (25:24:1, v/v) (Invitrogen, Waltham, MA, USA).

2.1.2 In Vitro Transcription

1. DNA/RNA LoBind Tubes, 1.5 mL (Eppendorf, Hamburg, Germany).
2. Diethyl pyrocarbonate (DEPC)-treated water (*see* **Note 2**).
3. Transcription mixture components: 10× In vitro Transcription (IVT) buffer (500 mM Tris–HCl, pH 7.9, 250 mM $MgCl_2$, 20 mM spermidine–HCl, 300 mM dithiothreitol (DTT)), 25 mM ribonucleotide triphosphate mix solution (ATP, CTP, UTP, and GTP), 0.1 U/μL yeast inorganic pyrophosphatase (New England Biolabs, Ipswich, MA, USA), and 20 U/μL phage T7 RNA polymerase (Thermo Fisher Scientific, Waltham, MA, USA) (*see* **Note 3**).

4. Amicon Ultra-15 centrifugal filtration devices with a molecular weight cutoff of 10,000 Da (MilliporeSigma, St. Louis, MO, USA).

2.1.3 Polyacrylamide Gel Electrophoresis (PAGE)

1. RNaseZap (Invitrogen, Waltham, MA, USA).
2. 10× Tris/Borate/EDTA (TBE) buffer: 890 mM Tris-borate, pH 8.3, 20 mM EDTA.
3. 2 L 40% acrylamide–bis-acrylamide stock solution 29:1 (w/w): 27.5 g N,N′-methylenebisacrylamide and 772.4 g acrylamide (MilliporeSigma, St. Louis, MO, USA) dissolved in 2 L of DEPC water. Store protected from light at 4 °C.
4. Polyacrylamide gel (PAG) stock solution: 1× TBE buffer, 8 M urea (MilliporeSigma, St. Louis, MO), 20% acrylamide–bisacrylamide solution 29:1 (w/w). Store in the dark for up to 3 months.
5. TBE/urea solution: 1× TBE, 8 M urea solution. Store in the dark for up to 3 months.
6. Denaturing loading buffer: 1× TBE, 8 M urea, 0.03% (w/v) xylene cyanol and 0.03% (w/v) bromophenol blue.
7. 10% ammonium persulfate (APS) solution. Store at 4 °C in the dark for no more than 4 weeks.
8. N,N,N′,N′-Tetramethylethylenediamine (TEMED).
9. Mega-Gel Dual High-Throughput Vertical electrophoresis gel system (C.B.S. Scientific, San Diego, CA, USA). Custom glass plates measuring 45 × 50 cm and accessories: three 0.5 cm spacers, and a single-tooth comb (45 cm width).
10. Heat dissipation metal plate (similar glass plates dimensions).
11. Silica gel thin layer chromatography (TLC) plates, 20 × 20 cm (Selecto Scientific, Suwanee, GA, USA).
12. Handheld UV lamp (365 nm).
13. Disposable scalpels.
14. Elutrap electroelution systems, with four Elutrap chambers. BT1 and BT2 membranes for electroelution (Cytiva, Marlborough, MA, USA).
15. Electroelution buffer: 20 mM Tris-acetate, pH 8.0.
16. Ethanol precipitation reagents: 3 M sodium acetate, pH 5.2, 100% ethanol, 80% (v/v) ethanol solution.
17. Nanodrop or UV spectrophotometer.

2.1.4 RNA Chromatography

1. Mono Q 5/50 (1 mL) column (Cytiva, Marlborough, MA, USA).
2. AKTA pure chromatography system (Cytiva, Marlborough, MA, USA).

3. Column cleaning solutions: 0.5 M NaOH, 2 M NaCl.
4. Buffer A: 20 mM Tris–HCl, pH 7.0.
5. Buffer B: 20 mM Tris–HCl, pH 7.0, 1 M NaCl.
6. Column storage solution: 20% ethanol.
7. 0.22 μm 25 mm syringe filters
8. 10 mL syringes.

2.2 Crystallization

1. RNA folding/binding buffer F1 (for the RNA obtained from the thiM-HDV construct): 50 mM potassium acetate, pH 6.9, and 5 mM $MgCl_2$.
2. RNA folding/binding buffer F2 (for the RNA obtained from the HH-thiM-VS construct): 5 mM Tris-HCl, pH 8.0, 3 mM $MgCl_2$, 10 mM NaCl, 100 mM KCl, and 0.5 mM spermine.
3. Reservoir solution A (for the TPP–riboswitch complex prepared with the thiM-HDV RNA): 100 mM Na-acetate, pH 4.8, 200 mM ammonium acetate, and 28% (w/v) polyethylene glycol (PEG) 4000.
4. Reservoir solution B (for the TPP–riboswitch and thiamine-riboswitch complexes prepared with the HH-thiM-VS RNA): 50 mM BisTris-HCl, pH 6.5, 500 mM NH_4Cl, 10 mM $MnCl_2$, and 30% (w/v) PEG2000.
5. Reservoir solution C (for the HMP-riboswitch and PT-riboswitch complexes prepared with the HH-thiM-VS RNA): 50 mM sodium cacodylate, pH 6.0, 250 mM NH_4Cl, 10 mM $CaCl_2$, and 30% (w/v) PEG2000.
6. Cryoprotection solution: The same reservoir solution, with all components concentrations increased by 10% and additionally including 1 mM of the ligand and 15% of glycerol.
7. Pre-greased crystallization plates with 24 wells (Hampton Research, Aliso Viejo, CA, USA).
8. 18 or 22 mm siliconized glass cover slides, depending on the plate.
9. Compressed gas can for removing dust from wells and cover slides.
10. Stereomicroscope, 12:1 zoom range with polarizing filters, for example, Olympus SZX16.
11. CrystalCap SPINE HT. Magnetic Goniometer Base mounted with a cryoLoop (0.05–0.3 mm sizes) (Hampton Research, Aliso Viejo, CA, USA).
12. Cryogenic container for liquid nitrogen.

3 Methods

3.1 RNA Preparation

Production of sufficient amount of RNA is essential to carry out crystallization experiments. In vitro transcription, performed with the phage T7 RNA polymerase [52], is the best method for obtaining milligram quantities of RNA for structural studies [53]. However, this enzyme produces RNA molecules with additional nontemplate nucleotides on the 3′-terminus [54], and depending on the sequence of the template, can also introduce nontemplate nucleotides on the 5′-end [55, 56]. Since crystal packing interactions often involve the ends of RNA molecules, it is essential to prepare RNA with homogeneous ends for crystallization experiments. Preparative chromatography techniques, typically used for protein purification, do not allow distinguishing between RNA species longer than 50 nt with a nucleotide resolution even under denaturing conditions. Therefore, to produce longer RNAs, structural biologists typically use denaturing polyacrylamide gels. Although such gels can discriminate between longer RNAs with nucleotide resolution, one has to load relatively small quantities to successfully separate RNAs longer than 30 nt. Thus, production of large amounts of the homogeneous 83 nt long *E. coli thi*M riboswitch RNA could represent a significant challenge. One way to increase the amount of the RNA to be loaded on the gel is to add extra RNA sequences on the RNA termini, cleave these pieces from the desired RNA, and then remove the pieces by the gel. In this method, heterogeneous termini are removed with the extra RNAs. If the extra RNAs are sufficiently different in size from the desired RNA, a gel can resolve rather large quantities of loaded RNAs. Specific cleavage of RNA can be achieved by *cis* and *trans* acting ribozymes.

Homogeneous *E. coli thi*M riboswitch RNA can be prepared by using two published strategies based on utilization of different ribozymes: (i) riboswitch RNA sequence flanked with a hammerhead (HH) ribozyme and the *Neurospora* Varkund Satellite (VS) substrate stem-loop on the 5′- and 3′-ends, respectively (HH-thiM-VS) [45, 57–61]; and (ii) a hepatitis delta virus (HDV) ribozyme encoded on the 3′-end of the riboswitch sequence (thiM-HDV) (Fig. 3) [57, 61]. The HH ribozyme requires the sequence upstream of the cleavage site to be UX, X $\neq$ G and, to cleave in *cis*-, the ribozyme sequence has to be complementary to the terminus of the riboswitch (Fig. 3a). The VS ribozyme has minimal sequence requirements, cutting after any base other than C, and cleaving in *trans*- after any sequence. Thus, the former strategy yields the riboswitch RNA of 83 nt in length with a 5′-OH and a 2′,3′-cyclic phosphate on the 3′ end. The HDV ribozyme cuts efficiently after any base other than G (Fig. 3b). Therefore, the latter strategy generates the RNA product of 85 nt

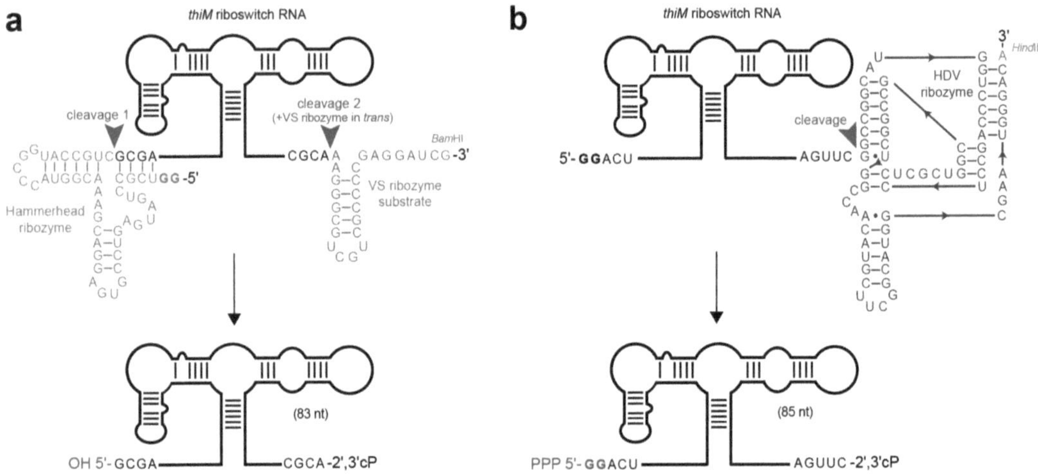

Fig. 3 Methods used for preparing homogeneous *thiM* TPP riboswitch RNAs for crystallization experiments. (**a**) Riboswitch RNA construct from Ref. [45]. The riboswitch is flanked with hammerhead (HH) ribozyme and Varkund satellite (VS) substrate (top panel). Riboswitch, HH and VS ribozymes are shown in black, blue, and green colors, respectively. Red arrows show ribozyme cleavage sites. A nucleotide left after cleavage of the DNA by *Bam*HI is in orange. Ribozyme cleavages, indicated by black arrows, produce an RNA of 83 nt in length with a 5′-OH and a 2′,3′-cyclic phosphate (2′,3′cP) on the 3′ end (bottom panel). (**b**) Riboswitch RNA construct from Ref. [44]. The riboswitch (black) is followed by a hepatitis delta virus (HDV) ribozyme (purple) (top panel). A nucleotide left after cleavage of the DNA by *Hind*III is in orange. The ribozyme cleavage produces a 85 nt in length with a 5′-triphosphate (PPP) and a 2′,3′cP on the 3′ end (bottom panel). Differences between RNA products in panels (**a**) and (**b**) are shown in red color. A pair of the 5′-terminal guanosines incorporated for improvement of transcription carried out by the T7 RNA polymerase is in bold

with a 5′-tri-phosphate and 2′,3′-cyclic phosphate on the 3′ end. This RNA also retains two Gs on the 5′ end to increase transcription efficiency. Note that the size of the final riboswitch RNA is sufficiently different from the size of other transcription and cleavage products to favor its purification by PAGE in large quantities.

3.1.1 DNA Template Preparation

The RNA constructs, cloned in high copy number plasmids, are driven by the T7 promoter and followed by a specific restriction site for runoff transcription: *Bam*HI in the pHH-thiM-VS plasmid encoding HH-thiM-VS RNA; and *Hind*III in the pthiM-HDV and pVSRZ encoding the thiM-HDV RNA and the VS *trans*-acting ribozyme, respectively (Fig. 3).

1. Grow bacterial strains with the desired riboswitch and ribozyme constructs on LB agar/Amp plates at 37 °C overnight.
2. For large-scale plasmid production, transfer individual bacterial clones into 4 L of LB/Amp medium and grow bacteria at 37 °C overnight. Extract and purify the plasmid DNA using a commercial plasmid extraction kit. The scale of the kit depends on the amount of the plasmid DNA required for experiments. The Gigaprep kit, typically used by us, yields around 6–10 mg of the plasmid DNA.

3. Prepare the DNA template for in vitro transcription. In the case of the HH-thiM-VS construct, linearize 1 mg of the pHH-thiM-VS with *Bam*HI-HF (2 U/μg plasmid), and 200 μg of the pVSRZ with *Hind*III-HF (2 U/μg plasmid), in the presence of 1× CutSmart buffer overnight at 37 °C. The thiM-HDV template is prepared by cleaving 1 mg of the pthiM-HDV construct with *Hind*III (2 U/μg plasmid) in 1× CutSmart buffer at 37 °C for ~4 h. It is recommended to verify the digestion completion using a 0.8% agarose gel electrophoresis in TAE 1X buffer. Continue reaction if the DNA plasmid has not been entirely digested.
4. Purify the linearized DNA templates using a phenol-chloroform extraction. Dilute the digestion product in 10 mL of water and add 1 volume of phenol:chloroform:isoamyl alcohol. Vortex the mixture for 1 min in a conical tube and centrifuge at 4000 × g for 5 min. Carefully collect the upper aqueous phase and repeat the extraction adding 1 volume of chloroform.
5. Precipitate the DNA template with 0.1 volume of 3 M Na-acetate, pH 5.2, and 3 volumes of 100% ethanol, in a 50 mL conical tube. Incubate the precipitation mixture at −20 °C overnight and centrifuge for 60 min at the maximal speed (>10,000 × g). Discard the supernatant and wash the pellet with ~10 mL of 80% ethanol and centrifuge for another 15 min. Gently air-dry the pellet (*see* **Note 4**). Dissolve the pellet in 2–5 mL of DEPC-treated water.
6. Determine the concentration of the DNA templates using a UV spectrophotometer at 260 nm wave length assuming 50 μg/mL DNA for 1 absorbance unit.
7. In the case of HH-thiM-VS construct, prepare a 1:10 mixture of linearized pVSRZ and pHH-thiM-VS plasmids. For example, combine 5 μg/mL of pVSRZ and 50 μg/mL of pHH-thiM-VS.

3.1.2 In Vitro Transcription

The in vitro transcription reaction must be optimized in a small-scale prior to performing a large-scale reaction. Transcription of each DNA template may need to be optimized by varying concentrations of DNA template (40–100 mg/mL), T7 RNA polymerase (30–100 μg/mL), and $MgCl_2$ (10–20 mM); and the incubation time (from 1–3 h up to overnight).

1. Small-scale in vitro transcription reaction: a typical reaction is carried out in a total volume of 50 μL, combining 2.5 μg linear DNA template (pthiM-HDV) or 2.5 μg premixed linear DNA templates (10:1, pHH-thiM-VS:pVSRZ), 5 μL 10× IVT buffer, 5 μL 25 mM NTP mix, 1 μL yeast inorganic pyrophosphatase (0.1 U/μL), and 40 U T7 RNA polymerase in DEPC-

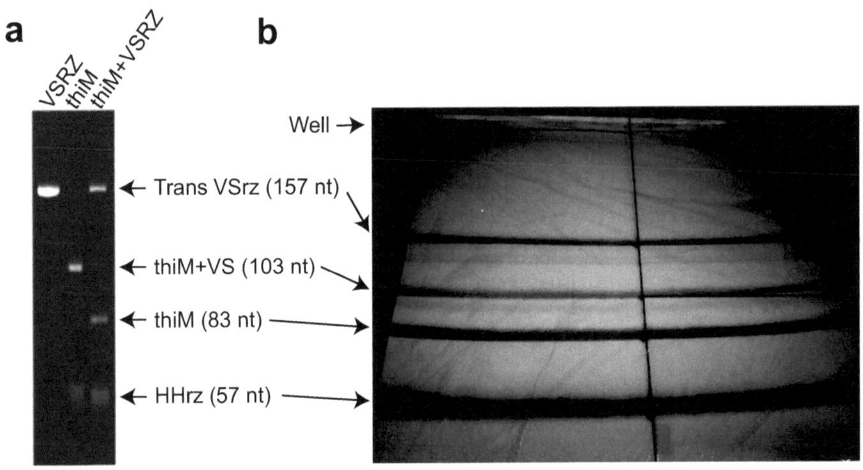

Fig. 4 PAGE of transcription and ribozyme cleavages of the HH-thiM-VS construct. (**a**) Small-scale and (**b**) large-scale transcription reactions were separated on analytical and preparative gels, respectively. (**a**) VSRZ lane, transcription of the Varkund satellite ribozyme (VSRZ) used for the *trans*-cleavage at the 3′ flank of the HH-thiM-VS construct. *thiM* and *thiM* + VSRZ lanes, transcription products obtained from the HH-thiM-VS construct in the absence or presence of VSRZ, respectively. Bands of the VS and HH ribozymes are shown as Trans VSrz and HHrz, respectively. RNA products after the HH and VS cleavages are shown as thiM+VS and thiM, respectively. The RNA lengths are in parentheses. The gel was stained by ethidium bromide. (**b**) RNAs on the preparative gel were visualized by the "UV shadow" method

treated water. Yeast inorganic pyrophosphatase can be omitted but it may increase transcription yield. Preheat the transcription mixture prior to the addition of DNA template(s) and the T7 RNA polymerase. Incubate the reaction for 3–6 h (continue overnight if necessary) at 37 °C. After the reaction, save 10 μL of the reaction for checking the RNA by PAGE. To enhance ribozyme cleavage, increase the final concentration of $MgCl_2$ to 50 mM and incubate for another 30 min at 37 °C. Save an aliquot for PAGE. Analyze transcription yield and efficiency of ribozyme cleavage by 10% small-scale PAGE. There are usually four bands after transcription using the pHH-thiM-VS:pVSRZ template mixture (Fig. 4a). The pthiM-HDV template typically yields three bands: ribozyme, riboswitch, and the transcript prior to the ribozyme cleavage.

2. Large-scale in vitro transcription: perform the reaction under optimized conditions in a total volume of 10 mL. For example, combine 500 μg DNA template, 1 mL 10× IVT buffer, 1 mL 25 mM NTP mix, 200 μL yeast inorganic pyrophosphatase (0.1 U/μL), and 15,000 U T7 RNA polymerase in DEPC-treated water. Preheat the mixture before adding the DNA template and T7 RNA polymerase. Add polymerase drop by drop to avoid aggregation. Incubate the transcription reaction for 4–6 h at 37 °C. Adjust the final concentration of $MgCl_2$ to 50 mM, and incubate for 30 min at 37 °C. The procedure can

be stopped at this point and the reaction can be stored at −20 °C. Verify the transcription yield and ribozyme cleavage by 10% small-scale PAGE.

3. Concentrate the transcription reaction using two Amicon Ultra-15 centrifugal filtration units. Centrifuge at 4000 rpm in a swinging bucket rotor at 4 °C. Reduce the volume to ~1 mL per device.
4. To reduce amount of NTPs and abortive transcripts, which can interfere with PAGE, add 10 mL DEPC-treated water to each filter and centrifuge at 4000 rpm until the volume is again reduced to ~1 mL per device. Dilute and concentrate the reaction three times, reducing the total volume of reaction to ~500 μL per device in the last concentration (~1 mL total). In the end, all buffer components will be reduced ~1000-fold.

3.1.3 Large-Scale PAGE for RNA

1. The concentrated transcription product is purified by preparative PAGE performed on a MegaGel Dual-Adjustable Vertical electrophoresis system using customized large glass plates (45 cm length × 50 cm width). The electrophoresis chambers, glass plates and comb must be scrubbed clean with RNaseZap or another RNase decontamination solution, rinse thoroughly with RNase-free water and 100% ethanol.
2. Gel casting: Position three spacers (0.5 cm thick) between both glass plates along the sides and bottom edges, and clamp the glass plates together. Prepare 1 L of 8% PAG/8 M urea 1× TBE solution by mixing 0.5 L of 20% PAG stock solution with 0.5 L of TBE/Urea solution. To avoid leaks during polymerization, prepare a plug by polymerizing small amount of PAG along spacers. Mix 100 mL of PAG/Urea/TBE solution with 600 μL of 10% APS and 40 μL of TEMED and pour along the side spacers. The acrylamide solution deposited on the bottom spacer must be in contact with the entire length of spacers to allow the gel to polymerize with glass plates. After polymerization of the plug, mix the remaining 900 mL of PAG/Urea/TBE solution with 5.4 mL of APS solution and 360 μL of TEMED, and pour on top of the plug, pouring from a side of top part of glass plates to avoid bubbles. Insert a comb and fix it with clamps. Allow the gel to polymerize at about 30° angle for 1 h.
3. Clean thoroughly the surface of the outer glass plate and clamp a metal plate to ensure the heat distribution along the gel during the electrophoresis.
4. Remove the bottom spacer and the comb, and clamp the gel in the Dual-Adjustable Mega-Gel system. Add 1 L of 1× TBE buffer to each electrophoresis chamber. Wash the well of the

gel with buffer and suck possible bubbles from the void left by the bottom spacer, with the help of a syringe and a curved needle.

5. Preheat the gel at 80 W for 1 h.
6. Stop the gel and load transcription reaction mixed with 1 mL of the 8 M urea-containing loading buffer. Run the gel at 80 W around 10 h until the first dye, bromophenol blue, exits the gel.
7. After the run, position the gel assembly on a clear bench and remove the top glass plate, being easier on the notched plate. Cover the gel with Saran Wrap, flip the glass plate, placing the gel onto a clean flat surface, and remove the second glass plate.
8. Carefully, place silica gel TLC plates under Saran Wrap and gel. Illuminate the gel with a UV lamp (preferably at 365 nm to avoid RNA damage) and locate the RNA as a dark band in the gel. This method of RNA visualization is called "UV shadow." Cut the RNA band out of the gel with a disposable scalpel. The gel slices can be stored at −20 °C. The TPP riboswitch band is well-separated from uncleaved and semi-cleaved RNAs, and ribozymes products (Fig. 4b) for both riboswitch constructs.
9. RNA is extracted from the gel slices by an Elutrap electroelution system. Using clean forceps, assemble an Elutrap chamber with the corresponding membranes. Place a BT1 trapezoidal impermeable membrane at each end of the chamber, between the two U-shaped inserts (the slanted tip faced to the arrow on the chamber); and a BT2 rectangular semipermeable membrane between the two U-shaped inserted closer to the middle of the chamber. BT2 divides the chamber in two sections: a large one for gel slices and a small one for the elution. Place the gel slices into the big section of the chamber, and place the chamber onto a tray of the Elutrap base. Fill the Elutrap chamber and base with the Electroelution buffer (20 mM Tris-acetate, pH 6.8) until the gel is completely covered. Close the Elutrap lid and perform electrophoresis at 5 W per Elutrap chamber used (20 W for the full system loaded) at 4 °C in a cold room.
10. Collect eluate every 2–3 h for the first 8 h and determine the RNA concentration at 260 nm by a UV spectrophotometer. If necessary, electrophoresis can continue overnight at 70 V.
11. Precipitate the RNA with 0.1 volume of 3 M Na-acetate, pH 5.2, and 3 volumes of 100% ethanol, in 50 mL conical tubes, following the instruction described in Subheading 3.1.1. Dry the pellet on air and dissolve in 200 μL of DEPC-treated water (*see* **Note 5**).

3.1.4 RNA Chromatography

Use anion-exchange chromatography on a Mono-Q column to purify the RNA from traces of urea and other contaminants. This step may not be required for robustly crystallizing riboswitch–ligand complexes. RNA must be filtered through a 0.2 μm 25 mm nitrocellulose syringe filter to remove possible particles of acrylamide.

1. The column must be cleaned before starting the chromatography. Wash column in 4 column volumes (CV) of 0.5 M NaOH, 2 CV of 2 M NaCl and followed by at least 2 CV of DEPC-treated water until baselines for eluent pH and conductivity are stable.
2. Balance the column with 5–10 CV of Buffer A.
3. Load slowly the sample by injection directly into the column, loading ~2–4 mg of the RNA per run.
4. Elute the RNA using a 10 CV 0–100% gradient of Buffer B. After this first elution, only half of the RNA is eluted. It is recommended to repeat the elution several times using a step gradient of 100% Buffer B.
5. Clean the column in water after chromatography and store it in 20% ethanol at 4 °C.
6. Precipitate the eluted RNA by ethanol precipitation as described in Subheading 3.1.1. Dry the pellet on air. Dissolve RNA in 200 μL of DEPC-treated water, and determine the RNA concentration by UV spectroscopy.

3.2 Crystallization

Although sometimes crystallization conditions have to be rescreened for successful crystallization of complexes between riboswitches and new ligands [45, 49, 50], we found that often a small optimization of established conditions is sufficient to get TPP riboswitch crystals of good quality with ligands other than TPP (Fig. 5). Typically, we vary precipitant, PEG2000, concentration with 1% increments, pH with 0.2 unit increments, and salt concentration with 50 mM increments. If crystals look small and originate from multiple crystallization initiation points (Fig. 5e, f), lowering precipitant or salt concentration usually reduces the number of crystals and increases their size. In some cases, changing divalent cations, for example, from Mg^{2+} to Mn^{2+}, or mixing two different cations could help to improve quality of the crystals. In addition, since the TPP riboswitch was crystallized in two different forms using two different RNAs, if the complex with one RNA does not crystallize, one could try another RNA. While the riboswitch from Ref. [44] was crystallized at higher resolution (2.0 Å with TPP), this RNA is sensitive to crystallization conditions and requires fine tuning of precipitant concentration. The riboswitch construct from Ref. [45] is more robust in crystallization trials and yields crystals

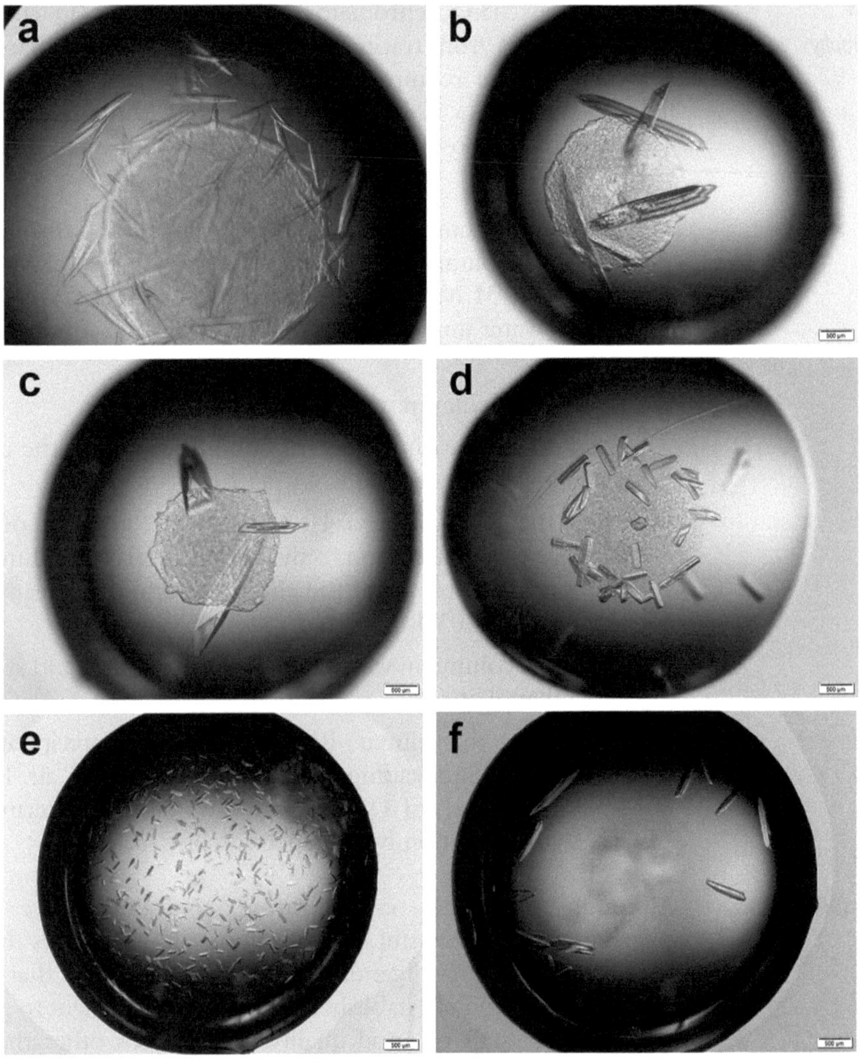

Fig. 5 Crystals of ligand–riboswitch complexes. (**a**, **b**) Crystals of the TPP–riboswitch complexes prepared with the RNA from Refs. [44, 45, 50] grew in solutions A and B, respectively. The bar scale is 500 μm. (**c**, **d**) Crystals of thiamine- and HMP–riboswitch complexes, respectively. (**e**) PT–riboswitch complexes grown in solution B produced small crystals with poor diffraction, whereas crystals set up in solution C produced bigger crystals with improved diffraction (**f**)

more readily; however, these crystals diffract to somewhat lower resolution, 2.2 Å with TPP and about 3.0 Å with other ligands. Lastly, the latter riboswitch was crystallized under three different sets of conditions [45, 50], and one has to try all of them because some conditions may produce small crystals (Fig. 5e) or poorly diffracting crystals with certain RNA–ligand complexes.

We typically set up co-crystallization experiments with the TPP riboswitch and its ligands by the hanging drop vapor diffusion method using pre-greased 24-well plates and glass cover slides.

Here, we describe crystallization procedures for two RNAs with TPP and for one RNA with three more ligands, thiamine, pyrithiamine, and HMP. All these complexes produce RNA structures with high overall similarity but with minor ligand-dependent conformational differences, likely important to accommodate a certain small molecule ligand.

1. Prepare the TPP riboswitch RNA in complex with its ligands by incubating 0.15 mM of the RNA with 0.5 mM of TPP, 1 mM thiamine, 1 mM of PT, or 2 mM HMP, in the RNA folding buffer F1 or F2, according to the RNA–ligand complex, for 30 min at 37 °C. The mixture could be stored for 1 h at 4 °C prior to crystallization.
2. Prepare 0.5 mL–1 mL of the reservoir solution and add it to the well of a 24-well plate (reservoir volume depends on the well volume of plates used for crystallization). We recommend to blow away dust particles from the wells just before adding the solution.
3. Take a round glass slide and remove dust particles from the surface using a compressed air can.
4. Put a drop of the riboswitch–ligand complex in the center of the glass slide. For the TPP–riboswitch complex with the HH-thiM-VS RNA, place a 1.5 μL drop of the riboswitch–ligand complex, and add 0.75 μL of the respective reservoir solution, B or C, to the drop of the complex. For the TPP–riboswitch complex prepared with the thiM-HDV RNA, mix the same volumes (1.5 μL) of the TPP–riboswitch complex and the reservoir solution A (1:1 ratio). In all cases, gently mix the final drop by pipetting.
5. Flip the glass slide, place it over the well, and press it down to make sure that the glass slide is fully sealing the well.
6. Place the plate in the incubator and incubate at 18 °C. Diffraction quality crystals of TPP (Fig. 5a, b), thiamine (Fig. 5c), and PT (Fig. 5e, f) in complex with the riboswitch appear in 1 week, while the HMP-riboswitch crystals grow over 3–4 weeks (Fig. 5d).
7. Prior to data collection, crystals need to be transferred into the cryoprotection solution. Find suitable crystals using a stereomicroscope. Pick a single crystal using a nylon loop of the appropriate size and pass the crystal through the 0.2–0.5 μL drop of the cryoprotecting solution. Flash-freeze a crystal by plunging into liquid nitrogen.

4 Notes

1. LB medium must be cooled to room temperature before adding ampicillin to avoid degradation of the antibiotic. In the case of LB/agar plates, make sure that the antibiotic is added before polymerization of the agar begins. At the appropriate temperature, the agar should not be too hot to the touch.
2. DEPC is used to prepare RNase-free solutions by irreversible modifications of catalytic residues of RNases. Prepare the DEPC-treated water by adding 1 mL of fresh DEPC to 2 L of bi-distillated water. DEPC is poorly soluble in water so it must be well mixed. Incubate overnight at room temperature to inactivate ribonucleases and autoclave the solution at 121 °C for 1 h for decomposing DEPC.
3. All materials and solutions used for RNA studies must be RNase-free. Prepare solutions in the DEPC-treated water and with RNase-free chemicals. Glassware can be cleaned from RNases by heating at 160 °C for at least 1 h.
4. Do not overdry DNA to ensure that it will be solubilized in water easily.
5. It is recommended but not required to check the RNA integrity by PAGE after its purification and extraction in the electroelution system. Nondegraded eluates can be combined.

References

1. Serganov A, Nudler E (2013) A decade of riboswitches. Cell 152:17–24
2. Mandal M, Breaker RR (2004) Gene regulation by riboswitches. Nat Rev Mol Cell Biol 5: 451–463
3. Nudler E, Mironov AS (2004) The riboswitch control of bacterial metabolism. Trends Biochem Sci 29:11–17
4. Winkler W, Nahvi A, Breaker RR (2002) Thiamine derivatives bind messenger RNAs directly to regulate bacterial gene expression. Nature 419:952–956
5. Roth A, Breaker RR (2009) The structural and functional diversity of metabolite-binding riboswitches. Annu Rev Biochem 78:305–334
6. Richards J, Belasco JG (2021) Widespread protection of RNA cleavage sites by a riboswitch aptamer that folds as a compact obstacle to scanning by RNase E. Mol Cell 81(127–138): e124
7. Peselis A, Gao A, Serganov A (2015) Cooperativity, allostery and synergism in ligand binding to riboswitches. Biochimie 117:100–109
8. Panchapakesan SSS, Breaker RR (2021) The case of the missing allosteric ribozymes. Nat Chem Biol 17(4):375–382
9. Serganov A, Patel DJ (2012) Metabolite recognition principles and molecular mechanisms underlying riboswitch function. Annu Rev Biophys 41:343–370
10. Baker JL, Sudarsan N, Weinberg Z et al (2012) Widespread genetic switches and toxicity resistance proteins for fluoride. Science 335: 233–235
11. Nahvi A, Sudarsan N, Ebert MS et al (2002) Genetic control by a metabolite binding mRNA. Chem Biol 9:1043
12. Blount KF, Breaker RR (2006) Riboswitches as antibacterial drug targets. Nat Biotechnol 24: 1558–1564
13. Blount KF, Megyola C, Plummer M et al (2015) Novel riboswitch-binding flavin analog that protects mice against Clostridium difficile infection without inhibiting cecal flora. Antimicrob Agents Chemother 59:5736–5746

14. Blount KF, Wang JX, Lim J et al (2007) Antibacterial lysine analogs that target lysine riboswitches. Nat Chem Biol 3:44–49
15. Serganov A, Huang L, Patel DJ (2008) Structural insights into amino acid binding and gene control by a lysine riboswitch. Nature 455: 1263–1267
16. Serganov A, Huang L, Patel DJ (2009) Coenzyme recognition and gene regulation by a flavin mononucleotide riboswitch. Nature 458:233–237
17. Mulhbacher J, Brouillette E, Allard M et al (2010) Novel riboswitch ligand analogs as selective inhibitors of guanine-related metabolic pathways. PLoS Pathog 6:e1000865
18. Ster C, Allard M, Boulanger S et al (2013) Experimental treatment of Staphylococcus aureus bovine intramammary infection using a guanine riboswitch ligand analog. J Dairy Sci 96:1000–1008
19. Yan LH, Le Roux A, Boyapelly K et al (2018) Purine analogs targeting the guanine riboswitch as potential antibiotics against Clostridioides difficile. Eur J Med Chem 143: 755–768
20. Pedrolli DB, Matern A, Wang J et al (2012) A highly specialized flavin mononucleotide riboswitch responds differently to similar ligands and confers roseoflavin resistance to Streptomyces davawensis. Nucleic Acids Res 40: 8662–8673
21. Lee ER, Blount KF, Breaker RR (2009) Roseoflavin is a natural antibacterial compound that binds to FMN riboswitches and regulates gene expression. RNA Biol 6:187–194
22. Sudarsan N, Cohen-Chalamish S, Nakamura S et al (2005) Thiamine pyrophosphate riboswitches are targets for the antimicrobial compound pyrithiamine. Chem Biol 12: 1325–1335
23. Howe JA, Wang H, Fischmann TO et al (2015) Selective small-molecule inhibition of an RNA structural element. Nature 526:672–677
24. Wang H, Mann PA, Xiao L et al (2017) Dual-targeting small-molecule inhibitors of the *Staphylococcus aureus* FMN riboswitch disrupt riboflavin homeostasis in an infectious setting. Cell Chem Biol 24:576–588. e576
25. Warner KD, Hajdin CE, Weeks KM (2018) Principles for targeting RNA with drug-like small molecules. Nat Rev Drug Discov 17: 547–558
26. Connelly CM, Moon MH, Schneekloth JS Jr (2016) The emerging role of RNA as a therapeutic target for small molecules. Cell Chem Biol 23:1077–1090
27. Mironov AS, Gusarov I, Rafikov R et al (2002) Sensing small molecules by nascent RNA: a mechanism to control transcription in bacteria. Cell 111:747–756
28. Barrick JE, Breaker RR (2007) The distributions, mechanisms, and structures of metabolite-binding riboswitches. Genome Biol 8:R239
29. Yadav S, Swati D, Chandrasekharan H (2015) Thiamine pyrophosphate riboswitch in some representative plant species: a bioinformatics study. J Comput Biol 22:1–9
30. Mukherjee S, Retwitzer MD, Barash D et al (2018) Phylogenomic and comparative analysis of the distribution and regulatory patterns of TPP riboswitches in fungi. Sci Rep 8:5563
31. Sudarsan N, Barrick JE, Breaker RR (2003) Metabolite-binding RNA domains are present in the genes of eukaryotes. RNA 9:644–647
32. Gong S, Du C, Wang Y (2020) Regulation of the thiamine pyrophosphate (TPP)-sensing riboswitch in NMT1 mRNA from Neurospora crassa. FEBS Lett 594:625–635
33. Shimizu M, Masuo S, Itoh E et al (2016) Thiamine synthesis regulates the fermentation mechanisms in the fungus Aspergillus nidulans. Biosci Biotechnol Biochem 80:1768–1775
34. Cheah MT, Wachter A, Sudarsan N et al (2007) Control of alternative RNA splicing and gene expression by eukaryotic riboswitches. Nature 447:497–500
35. Croft MT, Moulin M, Webb ME et al (2007) Thiamine biosynthesis in algae is regulated by riboswitches. Proc Natl Acad Sci USA 104: 20770–20775
36. Bocobza S, Adato A, Mandel T et al (2007) Riboswitch-dependent gene regulation and its evolution in the plant kingdom. Genes Dev 21: 2874–2879
37. Wachter A, Tunc-Ozdemir M, Grove BC et al (2007) Riboswitch control of gene expression in plants by splicing and alternative 3′ end processing of mRNAs. Plant Cell 19:3437–3450
38. Li S, Breaker RR (2013) Eukaryotic TPP riboswitch regulation of alternative splicing involving long-distance base pairing. Nucleic Acids Res 41:3022–3031
39. Kubodera T, Watanabe M, Yoshiuchi K et al (2003) Thiamine-regulated gene expression of *Aspergillus oryzae thiA* requires splicing of the intron containing a riboswitch-like domain in the 5′-UTR. FEBS Lett 555:516–520
40. Chauvier A, Picard-Jean F, Berger-Dancause JC et al (2017) Transcriptional pausing at the translation start site operates as a critical

checkpoint for riboswitch regulation. Nat Commun 8:13892

41. Rodionov DA, Vitreschak AG, Mironov AA et al (2002) Comparative genomics of thiamin biosynthesis in procaryotes. New genes and regulatory mechanisms. J Biol Chem 277: 48949–48959
42. Kulshina N, Edwards TE, Ferre-D'amare AR (2010) Thermodynamic analysis of ligand binding and ligand binding-induced tertiary structure formation by the thiamine pyrophosphate riboswitch. RNA 16:186–196
43. Rentmeister A, Mayer G, Kuhn N et al (2007) Conformational changes in the expression domain of the Escherichia coli thiM riboswitch. Nucleic Acids Res 35:3713–3722
44. Serganov A, Polonskaia A, Phan AT et al (2006) Structural basis for gene regulation by a thiamine pyrophosphate-sensing riboswitch. Nature 441:1167–1171
45. Edwards TE, Ferre-D'amare AR (2006) Crystal structures of the thi-box riboswitch bound to thiamine pyrophosphate analogs reveal adaptive RNA-small molecule recognition. Structure 14:1459–1468
46. Thore S, Leibundgut M, Ban N (2006) Structure of the eukaryotic thiamine pyrophosphate riboswitch with its regulatory ligand. Science 312:1208–1211
47. Meredith J., Zeller Ashok, Nuthanakanti Kelin, Li Jeffrey, Aubé Alexander, Serganov Kevin M., Weeks (2022) Subsite Ligand Recognition and Cooperativity in the TPP Riboswitch: Implications for Fragment-Linking in RNA Ligand Discovery. ACS Chemical Biology 17(2) 438–448
48. Thore S, Frick C, Ban N (2008) Structural basis of thiamine pyrophosphate analogues binding to the eukaryotic riboswitch. J Am Chem Soc 130:8116–8117
49. Warner KD, Ferre-D'amare AR (2014) Crystallographic analysis of TPP riboswitch binding by small-molecule ligands discovered through fragment-based drug discovery approaches. Methods Enzymol 549:221–233
50. Warner KD, Homan P, Weeks KM et al (2014) Validating fragment-based drug discovery for biological RNAs: lead fragments bind and remodel the TPP riboswitch specifically. Chem Biol 21:591–595
51. Antunes D, Jorge NN, Garcia De Souza Costa M et al (2019) Unraveling RNA dynamical behavior of TPP riboswitches: a comparison between Escherichia coli and Arabidopsis thaliana. Sci Rep 9:4197
52. Milligan JF, Groebe DR, Witherell GW et al (1987) Oligoribonucleotide synthesis using T7 RNA polymerase and synthetic DNA templates. Nucleic Acids Res 15:8783–8798
53. Pikovskaya O, Serganov AA, Polonskaia A et al (2009) Preparation and crystallization of riboswitch-ligand complexes. Methods Mol Biol 540:115–128
54. Rong M, Durbin RK, Mcallister WT (1998) Template strand switching by T7 RNA polymerase. J Biol Chem 273:10253–10260
55. Helm M, Brule H, Giege R et al (1999) More mistakes by T7 RNA polymerase at the 5′ ends of *in vitro*-transcribed RNAs. RNA 5:618–621
56. Pleiss JA, Derrick ML, Uhlenbeck OC (1998) T7 RNA polymerase produces 5′ end heterogeneity during *in vitro* transcription from certain templates. RNA 4:1313–1317
57. Ferre-D'amare AR, Doudna JA (1996) Use of *cis*- and *trans*-ribozymes to remove 5′ and 3′ heterogeneities from milligrams of *in vitro* transcribed RNA. Nucleic Acids Res 24: 977–978
58. Guo HC, Collins RA (1995) Efficient *trans*-cleavage of a stem-loop RNA substrate by a ribozyme derived from neurospora VS RNA. EMBO J 14:368–376
59. Martick M, Scott WG (2006) Tertiary contacts distant from the active site prime a ribozyme for catalysis. Cell 126:309–320
60. Price SR, Ito N, Oubridge C et al (1995) Crystallization of RNA-protein complexes. I. Methods for the large-scale preparation of RNA suitable for crystallographic studies. J Mol Biol 249:398–408
61. Walker SC, Avis JM, Conn GL (2003) General plasmids for producing RNA *in vitro* transcripts with homogeneous ends. Nucleic Acids Res 31:e82

Chapter 15

Preparation of RNA Microcrystals for Serial Femtosecond Crystallography Experiments

Jason R. Stagno

Abstract

Serial femtosecond crystallography (SFX) experiments using an X-ray free electron laser (XFEL) is a burgeoning method for time-resolved structural studies of biomacromolecules. As with any crystallography experiment, the most important component is quality sample preparation. Whereas dozens of SFX experiments, including batch crystallization methods, have been reported for proteins, very few have been reported for RNA. This chapter outlines standard procedures for preparing RNA microcrystalline samples suitable for SFX studies.

Key words RNA batch crystallization, Serial femtosecond crystallography, X-ray free electron laser

1 Introduction

RNA structures are vastly underrepresented in the Protein Data Bank, where RNA or RNA-containing complexes account for only a few percent of total entries. Moreover, most of those entries comprise simple hairpins, duplexes, or short stretches of single-stranded RNA complexed with protein, and only a small fraction represents unique structures of larger, functional, noncoding RNAs. Aside from the fact that RNA structure gets far less attention than its more popular protein counterpart, there are several contributing factors that make RNA structure determination difficult. RNA molecules, although replete with secondary and tertiary structure, tend to be intrinsically dynamic, a feature which is associated with this macromolecule's myriad of functions. This property makes NMR a useful method for probing RNA structure and dynamics; however, the majority of biologically relevant RNAs are too large for NMR studies. On the other hand, most such RNAs are too small for cryo-electron microscopy studies. As a result, X-ray crystallography has proven to be the most suitable method for high-resolution structure determination of RNA molecules.

Jienyu Ding et al. (eds.), *RNA Structure and Dynamics*, Methods in Molecular Biology, vol. 2568,
https://doi.org/10.1007/978-1-0716-2687-0_15,

However, RNA's structural flexibility and negatively charged surface can make these macromolecules notoriously difficult to crystallize. If suitable RNA crystals can be grown for single-crystal X-ray diffraction (XRD) studies, they tend to suffer from high solvent content and high mosaicity, resulting in poor diffraction quality and resolution, and the electron-rich phosphate backbone make crystals of RNA highly sensitive to radiation damage. The birth of SFX using an XFEL [1] presented a means to overcome many of the obstacles faced by RNA structural biologists. With beam intensities of more than ten orders of magnitude greater than synchrotron light sources [2], and femtosecond X-ray pulses, SFX allows for high-resolution data collection at room-temperature using nano/microcrystals by outrunning radiation damage. XFEL-based methods also allow for time-resolved diffraction experiments ranging from picoseconds to seconds [3–5], generating a new era of time-resolved structural biology.

This chapter focuses on the preparation of microcrystalline sample of RNA, namely the adenine riboswitch aptamer domain (riboA), for its application in SFX studies. Apo structures of riboswitches have been extremely limited, most likely owing to difficulty in crystallization due to conformational flexibility in the absence of ligand. In our hands, however, we have found that several ligand-free riboswitch aptamer domains have the propensity to form nano/microcrystals that could potentially be used directly in XFEL-based studies or as seed stocks for attempts at obtaining larger crystals for conventional XRD studies (Fig. 1). For proteins and RNA alike, there may be cases in which solubility and thermodynamics during crystallization allow for molecular packing and crystal nucleation, but not long-range order. The identification and use of such crystals often rely on a combination of detection capabilities, namely birefringence, UV fluorescence (UVF), UV absorption (UVA), and second harmonic generation (SHG).

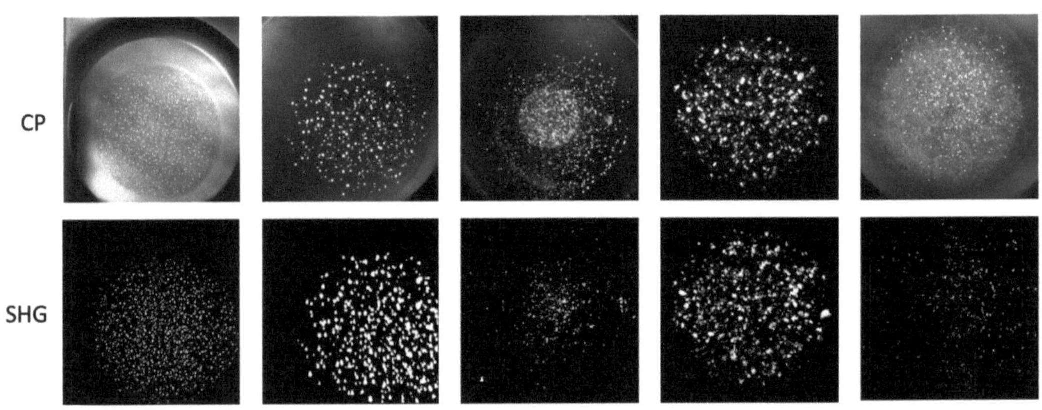

Fig. 1 Microcrystals of five different riboswitch aptamers, visualized under brightfield with cross-polarized light (CP) or by second harmonic generation (SHG) using second-order nonlinear imaging of chiral crystals (SONICC)

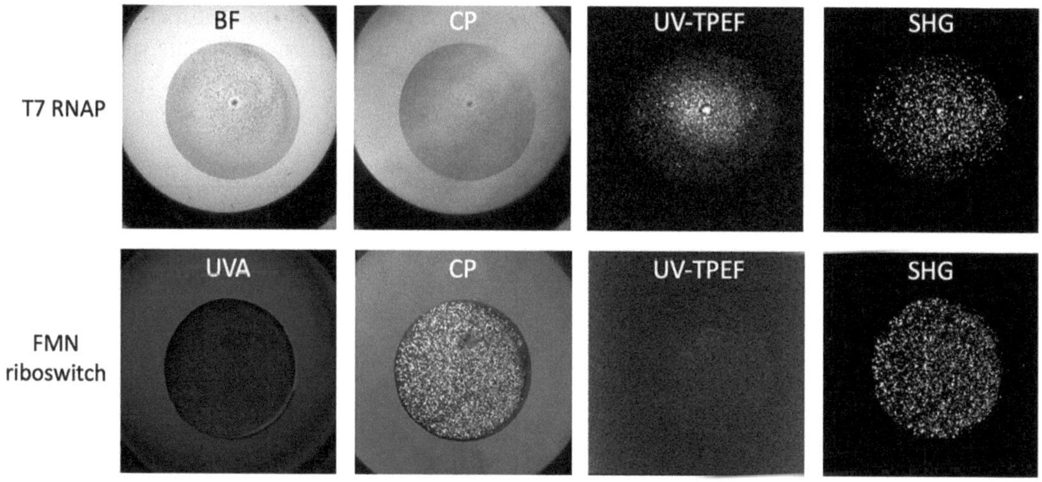

Fig. 2 Crystals of T7 RNA polymerase (top) and FMN riboswitch aptamer domain (bottom), imaged using bright-field (BF), bright-field with cross-polarization (CP), UV two-photon emission fluorescence (UV-TPEF), UV absorption (UVA), or second harmonic generation (SHG)

Crystal birefringence can be detected using a standard stereo microscope equipped with cross-polarization. RNA crystals tend to be very birefringent, making this a robust method for crystal detection, even for microcrystals. UV-based detection methods (UVF, UVA, UV two-photon emission fluorescence (UV-TPEF)) are also useful, particularly for distinguishing macromolecular crystals from salt crystals. In our hands, however, we have found detection by UVF/UV-TPEF to be less sensitive and to give ambiguous results in many cases, most likely due to the very high UV absorption of RNA, in which uncrystallized molecules in the surrounding mother liquor generate significant background noise. For RNA crystals, therefore, UVA is preferred for confirmation of RNA material, whereas birefringence is far more sensitive for detection (Fig. 2). The third, most recent and advanced detection method, second-order nonlinear imaging of chiral crystals (SONICC), detects even small levels of SHG signals propagated by chiral crystals, such as those of macromolecules (Figs. 1 and 2). This allows for both the detection of crystals as small as hundreds of nanometers, and the distinguishment from salt crystals, as most salt crystals are achiral [6]. Although advanced detection methods such as UV-TPEF and SONICC can be highly advantageous, they are quite expensive, and may not be practical for a single laboratory. However, an automated plate imager equipped with SONICC and a 1000-plate capacity (*e.g.*, RockImager, Formulatrix, Inc.) could be purchased jointly among multiple crystallography labs within the same department or institute for shared use, making this a much more viable option. Whereas bright-field (BF) imaging with cross-polarized (CP) light is sufficient for detection of microcrystals in most cases, the addition of SONICC can help ensure that no useful crystallization condition is overlooked. For example, microcrystals of T7 RNA polymerase were detectable by UV-TPEF and SONICC, but not by visible light (Fig. 2)

2 Materials

All buffers should be prepared at room temperature using ultrapure water that has been treated with 0.1% v/v diethylpyrocarbonate (DEPC) and autoclaved twice for 40 min each (DEPC-water). Filter all buffers through a 0.22 μm sterile filter.

1. RNA buffer: the final buffer in which the RNA is kept after purification, and which is used to dilute the RNA to desired concentrations.
2. Crystallization buffer: the mixture of precipitants whose concentrations have been optimized to generate a maximum density of RNA microcrystals under batch conditions.
3. Stabilization buffer: a mixture of crystallization buffer and RNA buffer at the same ratio used for the batch crystallization experiment. In this way, the RNA buffer is used in place of RNA to replicate the mother liquor as much as possible. For example: if the batch crystallization consists of equal parts crystallization buffer and RNA, then the stabilization buffer should consist of equal parts crystallization buffer and RNA buffer.
4. RNase-free 0.5 mm stainless steel beads (Next Advance, Inc., Troy, NY).
5. Rotary mixer.
6. UV-transmissible microplates.
7. Reusable glass cell-counting chamber.
8. Stereo microscope with cross-polarization.
9. Plate imager with UV and/or SONICC capabilities (optional).

3 Methods

Prepare RNA samples by in vitro transcription and purify them by native polyacrylamide gel electrophoresis (PAGE). Buffer-exchange purified RNA samples to a target RNA buffer in which the sample is stable (*e.g.*, 10 mM HEPES pH 7.5, 100 mM KCl, 0.1 mM EDTA, 0–10 mM $MgCl_2$). Concentrate the RNA to >10 g/L, treat with an RNase inhibitor (*e.g.*, SUPERaseIn^{TM}, Ambion, Inc. Austin, TX), and store the sample at −80 °C.

3.1 Initial Growth and Characterization of RNA Microcrystals

1. Establish native crystallization conditions by conventional vapor-diffusion or microbatch under-oil crystallization methods using UV-transmissible microplates. Crystal growth should be monitored using a combination of imaging techniques, including bright-field (BF), visible light with cross-polarization (CP), UV two-photon emission fluorescence (UV-TPEF), and second-harmonic generation (SHG) by SONICC (Fig. 1) (*see* **Note 1**).
2. If possible, initial crystallization conditions should be optimized to achieve dense showers of microcrystals.
3. Under optimized conditions, scale up the experiment in multiple crystallization plates that amounts to a minimum crystalline sample volume of 50 μL.
4. Harvest crystallization drops using 1–2 volumes of well buffer and transfer to a 1.5-mL microcentrifuge tube on ice.
5. Add 10–20 RNase-free stainless steel 0.5 mm beads.
6. Vortex for 30 s, then cool on ice for 1 min. Repeat a minimum of three times (*see* **Note 2**).
7. Make serial dilutions (*e.g.*, 1:10, 1:100, 1:1000) of the seed stock using the same crystallization buffer. Keep all seeds on ice and store at −80 °C when not in use. If **steps** 1–7 were done using the microbatch under-oil method, proceed to Subheading 3.2.
8. If the crystals have been grown thus far by vapor diffusion, it is highly recommended to establish similar conditions using microbatch under-oil, as the latter is much more suitable for scaling up to larger batch volumes (*see* **Note 3**). Repeat optimization procedure in **step** 2 using the microbatch under-oil method with or without the addition 1/10 V of seed stocks generated in **step** 7. As a starting point, one can create a 2D optimization grid that varies [RNA] and [precipitant] from 100 to 180% of the concentrations used for vapor diffusion.

3.2 Batch Crystallization Trials

1. Once a range of optimized microbatch crystallization conditions is established, repeat these conditions with various seed concentrations in sealed microcentrifuge tubes at volumes of 20–50 μL, without oil. Two methods can be tried here: (1) complete mixing to homogeneity by pipetting, followed by continued mixing using a rotary mixer; or (2) free-interface diffusion (FID), in which the reagent (RNA or precipitant/seeds) with higher density is slowly pipetted on top of the other to create a liquid–liquid interface (no induced mixing). The FID method may be useful particularly when the densities of the precipitants (*e.g.*, PEG, MPD) are significantly different from one another.

2. Optimize the conditions to produce dense showers of microcrystals, ideally within a few hours or days of incubation at room temperature.
3. Combine optimized crystalline samples and establish a new set of seed stocks (Subheading 3.1, **steps** 5–7).
4. Repeat **steps** 1 and 2 at larger volumes (*e.g.*, 0.2–0.4 mL) using both original and newly generated seed stocks. Here, the minimum volume must be such that the sample does not remain stationary when the tube is inverted during rotary mixing.
5. Incubate the solutions at room temperature with continual rotation (unless using FID).
6. Monitor crystal growth immediately upon mixing and periodically over several hours (or days) using a stereomicroscope with cross-polarized light until a maximum density (crystals/μL) of microcrystals has been achieved (*see* **Note 4**).
7. Identify the optimal crystallization conditions, including the seed stock and dilution used (*see* **Note 5**).

3.3 Optimizing Conditions for Large-Scale Batch Crystallization

1. Transfer 0.2–0.5 mL RNA to each of four 1.5-mL microcentrifuge tubes.
2. Add an appropriate amount of crystallization buffer to each tube per previously established microbatch crystallization conditions (*see* **Note 6**). Mix thoroughly by pipetting (*see* **Note** 7).
3. To the first tube, add nothing. To the next tube, add 1% v/v of original seed stock. To another tube, add 1% v/v of 1:10-diluted seed stock. To the last tube, add 1% v/v of 1:100-diluted seed stock.
4. Vary conditions to screen 1 or 2 parameters at a time (seed concentration, RNA concentration, precipitant concentration). Various combinations of these three parameters are likely to yield different outcomes with respect to crystal size, crystal density, crystal morphology, and the amount of precipitation (if applicable). In the case of riboA, all the RNA was either crystalline or granular precipitate. Therefore, optimizing the initial parameters was essential for maximizing crystal growth density while also minimizing precipitation. For example, as compared to 5 g/L RNA (with or without seeding), use of 1 g/L RNA and 0.1% v/v seed concentration without changing precipitant concentration yielded smaller and more uniform crystals with minimal precipitation (Fig. 3). Although the crystal density is approximately halved at the lower RNA concentration, the amount of RNA used is 80% less, thus yielding 2.5× as many crystals at more optimal size and uniformity, with minimal precipitate. In our experience, RNA precipitated from MPD tends to be highly granular, and does not

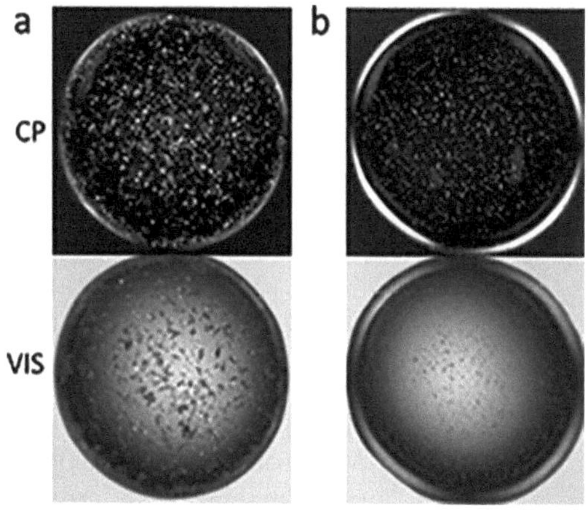

Fig. 3 Effects of seeding on batch crystallization of riboA. (**a**) Microcrystals of riboA suitable for SFX studies require a minimum of 5 g/L RNA. At this concentration, with or without seeds, the crystal sizes are nonuniform, most ranging from 2 to 10 μm (average of 5 μm), with a small population >10 μm. Crystals are clearly detectable under cross-polarized light (top), with crystal density ranging from 10^8 to 10^9 crystals per mL, depending on seeding volumes. When viewed without cross-polarization (bottom), however, this condition exhibits large amounts of precipitate, in which many of the crystals are imbedded. This precipitate is difficult to remove without significant loss of crystal matter, but if not filtered, can cause downstream issues with sample injection. (**b**) After optimizing batch crystallization conditions, the crystals are relatively uniform in size (1–2 μm) (top), and exhibit very minimal precipitate (bottom). Although crystal density is 2–3× less than the original condition, the crystals are generated using 80% less RNA, only a fraction of which is lost in the form of precipitate. Higher crystal density can be achieved simply by increasing the crystallization volume followed by gravity sedimentation

filter well during crystalline sample filtering steps. Therefore, high amounts of precipitate during crystallization will reduce the ratio of crystal: precipitate in the concentrated slurry, yielding lower hit-rates and increasing the chances of nozzle clogging during sample injection.

5. Repeat the procedure in **steps** 1–3 using larger (*e.g.*, 10×) volumes to ensure optimized conditions produce the desired results before proceeding to large-scale crystallization.

3.4 Large-Scale Crystallization of RNA Using the Batch Method

1. Thaw on ice the appropriate amount of purified RNA and crystal seed-stock to achieve the desired volume of >10^{10} crystals/mL for serial crystallography experiments. RNA and seed concentrations should match the optimal conditions established in Subheading 3.3.
2. Vortex the seed-stock for 30 s, then place it back on ice.
3. Centrifuge the RNA at >15,000 g for 10 min to remove particulate matter. Transfer supernatant to a fresh tube.

4. Split the RNA into sterile 50-mL conical tubes, with 10–20 mL per tube (*see* **Note 8**).
5. Add an appropriate amount of crystallization buffer to each tube per previously established crystallization conditions from Subheading 3.3.
6. Mix thoroughly by inverting the tubes several times.
7. Add an appropriate amount of seeds to each tube per previously established crystallization conditions from Subheading 3.3.
8. Incubate the solutions at room temperature with continual rocking or rotation (*see* **Note 9**).
9. Monitor crystal growth periodically using a stereomicroscope until maximum microcrystal density has been achieved (*see* **Note 10**).
10. Sediment the crystals in the 50-mL conical tubes (*see* **Note 11**).
11. Decant the supernatant into a sterile container (*see* **Note 12**).
12. Add 20 mL of stabilization buffer to each pellet and swirl the tube gently. Repeat **steps** 10–12 twice (*see* **Note 13**).
13. Ensure there is residual buffer in the tubes after decanting. If not, approximate the volume of the pellet and add an equal volume of stabilization buffer. Using a pipette, very gently resuspend the crystal pellets to achieve a dense slurry (*see* **Note 14**).
14. Combine the crystal slurry in a fresh sterile tube of appropriate size.
15. Make serial dilutions (10 μL each) of the crystal slurry using stabilization buffer.
16. Using a stereo microscope and a cell-counting chamber, identify a crystal dilution that gives a crystal density of 10–20 crystals per 1 μL volume. Count the crystals in the chamber and estimate the concentration of the crystal slurry (optimally $>10^{10}$ crystals/mL).

4 Notes

1. Use of SONICC can be particularly important if conventional screening generates nano/microcrystals that are undetectable under visible light.
2. To achieve a more homogenous seed stock, particularly if the crystals are difficult to fragment, the seeds can be passed through a 0.5 μm stainless steel frit after vortexing.
3. If suitable crystallization conditions cannot be established using the microbatch under-oil method, it is unlikely that the

crystals can be grown in larger batch volumes. In these cases, the crystals must be grown by vapor-diffusion by setting up numerous plates and harvesting the crystals drop by drop.

4. Crystal growth density can be estimated from droplets of samples on glass-coverslips or using a cell-counting chamber for more accurate measurements. With seeds, crystals generally grow faster. If further seeding optimization trials are necessary, repeat **steps** 1–6 using different seed concentrations and amounts and/or varying the RNA and precipitant concentrations.
5. Using the same seed stock and concentration is often important for consistent batch crystallization. If necessary, more of the seed stock should be generated following the same established protocol to minimize variations in seed stocks used for future experiments that may require much larger volumes.
6. Crystal growth, including size, density, and morphology, may vary with different batch volumes. If established conditions cannot generate the desired crystals with or without seeds at larger volumes, further optimization of RNA and/or precipitant concentrations may be required at this stage.
7. Mixing prior to the addition of seeds is critical to prevent dissolution of seeds.
8. A single experiment of 10–20 mL can be tried first to ensure desired results are achieved as expected in the larger volume.
9. Alternatively, crystals can be grown with gentle stirring with a magnetic stir-bar and stir-plate in a 50-mL conical tube or beaker. For large volumes in a single batch, larger beakers can be used.
10. Incubation times longer than necessary may result in crystals that are too large. The incubation time and seeding conditions may need to be optimized to maximize crystal density without overgrown crystals. Usually, this can be achieved with higher seed concentrations and/or slightly higher precipitant concentrations.
11. Although gravity sedimentation is recommended to avoid damaging the crystals, it is common practice to use gentle centrifugation (*e.g.*, 100 g). The speed of centrifugation should be as low as possible to achieve a soft pellet. Crystals with higher solvent content and buffers with higher viscosity may require higher speeds or longer centrifugation. This, however, is at the risk of damaging the crystals. Ideally, crystals with or without centrifugation should be tested by diffraction to determine the effects, if any.

12. If the mother liquor is not likely to interfere with downstream experiments (*e.g.*, mixing with ligand), it may be preferable to skip the washing of crystals with stabilization buffer (**Step** 12).
13. If a significant fractional volume of the sample is precipitate, one might consider filtering the sample through a 30–50 μm stainless steel frit (IDEX Health & Science, LLC, Oak Harbor, WA).
14. The crystal slurry should be easy to pipette after resuspension. If it is too thick or pasty, add more stabilization buffer.

Acknowledgment

This research was supported by the Intramural Research Program of the NIH, National Cancer Institute, Center for Cancer Research (Y-X. Wang).

References

1. Boutet S, Lomb L, Williams GJ et al (2012) High-resolution protein structure determination by serial femtosecond crystallography. Science 337:362–364
2. Huang N, Deng H, Liu B et al (2021) Features and futures of X-ray free-electron lasers. Innovation (NY) 2:100097
3. Pande K, Hutchison CDM, Groenhof G et al (2016) Femtosecond structural dynamics drives the trans/cis isomerization in photoactive yellow protein. Science 352:725–729
4. Ramakrishnan S, Stagno JR, Conrad CE et al (2021) Synchronous RNA conformational changes trigger ordered phase transitions in crystals. Nat Commun 12:1762
5. Stagno JR, Liu Y, Bhandari YR et al (2017) Structures of riboswitch RNA reaction states by mix-and-inject XFEL serial crystallography. Nature 541:242–246
6. Closser RG, Gualtieri EJ, Newman JA et al (2013) Characterization of salt interferences in second-harmonic generation detection of protein crystals. J Appl Crystallogr 46:1903–1906

Chapter 16

Mix-and-Inject Serial Femtosecond Crystallography to Capture RNA Riboswitch Intermediates

Jason R. Stagno, Juraj Knoska, Henry N. Chapman, and Yun-Xing Wang

Abstract

Time-resolved structure determination of macromolecular conformations and ligand-bound intermediates is extremely challenging, particularly for RNA. With rapid technological advances in both microfluidic liquid injection and X-ray free electron lasers (XFEL), a new frontier has emerged in time-resolved crystallography whereby crystals can be mixed with ligand and then probed with X-rays (mix-and-inject) in real time and at room temperature. This chapter outlines the basic setup and procedures for mix-and-inject experiments for recording time-resolved crystallographic data of riboswitch RNA reaction states using serial femtosecond crystallography (SFX) and an XFEL.

Key words Time-resolved crystallography, Serial femtosecond crystallography, X-ray free electron laser, RNA conformational intermediates, Diffusive mixing

1 Introduction

Throughout most of the history of X-ray crystallography, structure determination of conformational or reaction intermediates has been limited to proteins, primarily by soaking and freezing crystals to capture them in a particular state. In the case of enzymes, transition states can be achieved by this method using substrate analogs. Capturing structural changes in real-time, however, has been limited mostly to photoactive proteins using pump-probe Laue diffraction methods [1–3]. But light-induced reactions account for only a small fraction of molecular biology, leaving a huge gap in knowledge of biomolecular conformations induced by ligands, particularly at room temperature and in real time. Riboswitches are a prime example of RNAs with complex folds and structure–function-based roles, whose conformational trajectories are largely induced by binding of small-molecule ligands to regulate

Jienyu Ding et al. (eds.), *RNA Structure and Dynamics*, Methods in Molecular Biology, vol. 2568,
https://doi.org/10.1007/978-1-0716-2687-0_16,

gene expression. We used microcrystals of the adenine riboswitch aptamer domain (riboA) to carry out time-resolved diffraction studies using the mix-and-inject method [4–6].

This chapter briefly outlines the basic setup and procedures necessary to perform a mix-and-inject experiment. It does not, however, provide detailed procedures for the processing of serial crystallographic data. Regarding structure determination of time-resolved intermediates, this chapter outlines the use of sorted SFX data acquired for various time delays to calculate difference electron density (DED) maps between related data sets. In the cases of pump-probe experiments with, *e.g.*, photoactive proteins, where the light-induced molecular changes are mostly local, nondisruptive, and uniform throughout each crystal, singular value decomposition (SVD) methods can be applied to maximize signal-to-noise [2, 7, 8]. For large conformational changes induced by ligand diffusion, however, which are more heterogeneous both within and among crystals at acquired time delays, SVD methods may be less reliable, and are, therefore, not covered here.

2 Materials

All buffers should be prepared at room temperature using ultrapure water that has been treated with 0.1% v/v diethylpyrocarbonate (DEPC) and autoclaved twice for 40 min each (DEPC-water). Filter all buffers through a 0.22 μm sterile filter.

1. DEPC-water.
2. Stabilization buffer: a mixture of crystallization buffer and RNA buffer at the same ratio used for the batch crystallization experiment. In this way, the RNA buffer is used in place of RNA to replicate the mother liquor as much as possible. For example: if the batch crystallization consists of equal parts crystallization buffer and RNA, then the stabilization buffer should consist of equal parts crystallization buffer and RNA buffer.
3. Ligand buffer: stabilization buffer containing ligand at the desired concentration.
4. High-density (>10^9 crystals/mL) slurry of RNA microcrystals in stabilization buffer.
5. Pre-column PEEK filter housing (IDEX Health & Science, LLC, Oak Harbor, WA).
6. Stainless-steel frits, 5–100 μm (IDEX Health & Science, LLC, Oak Harbor, WA).
7. Sterile nylon mesh filters, 5-100 μm (*e.g.*, CellTrics®, Sysmex America, Inc., Lincolnshire, IL).

8. 1/16″ outer diameter PEEK tubing: inner diameters 75, 150, 250 μm (IDEX Health & Science, LLC, Oak Harbor, WA).
9. Luer adapters, finger-tight coned fittings, finger-tight coned ferrules for 1/16″ outer diameter PEEK tubing (IDEX Health & Science, LLC, Oak Harbor, WA).
10. Sample reservoir assemblies for liquid injection: 1-mL (Neptune Fluid Flow Systems, LLC, Knoxville, TN) and 10-mL (VariLoop S, KNAUER Wissenschaftliche Geräte GmbH, Berlin, Germany) reservoir volumes.
11. Plastic tubing cutter.
12. HPLC pumps (x2).
13. HPLC switchyards (x2), each with two high-pressure dead-end flowpath selectors with automated actuators.
14. High-precision 3D-printed double-flow focusing nozzle (DFFN) [9] or similar gas dynamic virtual nozzle (GDVN) (*see* **Note 1**).
15. High-precision 3D-printed microfluidic mixer (micromixer) [9] (*see* **Note 2**).
16. Anti-settling device.
17. Flow meter.
18. Helium gas for focusing of liquid jet.
19. Laboratory-grade ethanol for double-flow focusing of liquid jet.

3 Methods

3.1 Hardware Preparation for Mix-and-Inject Experiment

1. Soak all components of the sample reservoir assemblies in DEPC-water for 1 hr. These include but are not limited to stainless-steel sample holders and housings, Teflon plugs and ferrules, PEEK fittings, and adaptors.
2. Install micromixer and DFFN (*see* **Note 3**).
3. Prerinse all PEEK tubing and flush all HPLC and switchyard ports with DEPC-water.
4. Fill two 10-mL sample reservoirs with DEPC-water, two with stabilization buffer, and one with ligand buffer. Ensure no air remains in the reservoirs or connecting tubing.
5. Connect one of the DEPC-water reservoirs to the HPLC 1-controlled switchyard 1, and the other DEPC-water reservoir to the HPLC 2-controlled switchyard 2. Ensure no air is introduced into the system during the connections. Repeat for the two reservoirs containing stabilization buffer (Fig. 1) (*see* **Note 4**).

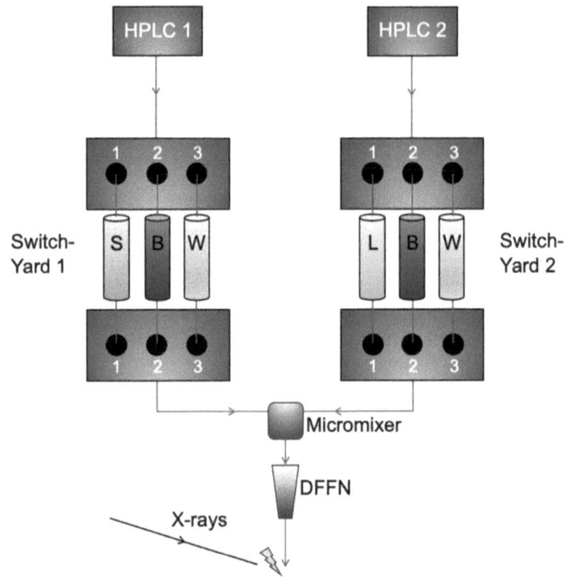

Fig. 1 Simplified schematic of mix-and-inject setup. Two HPLC pumps are used to regulate the flow of two independent liquid streams. The contents delivered to the micromixer are selected via two switchyards containing multiport valves with automated actuators. Switchyard 1 (left) contains the sample reservoir (S) and switchyard 2 (right) contains the ligand reservoir (L). Both switchyards also contain reservoirs filled with stabilization buffer (B) or DEPC-water (W), which are used for easy washing or unclogging

6. Connect the ligand-containing reservoir to switchyard 2.
7. Carefully resuspend the microcrystalline RNA sample and filter through a stainless-steel frit (10–30 μm, IDEX Health & Science, LLC, Oak Harbor, WA) into a fresh tube (*see* **Note 5**).
8. Fill a sample reservoir with the filtered sample and expel all air from the lines.
9. Connect the reservoir containing the microcrystalline sample to switchyard 1, again ensuring no air is introduced to the system, and attach it to the anti-settler (*see* **Note 6**).
10. Ensure both ends of each reservoir are connected to matching numbered ports on the flowpath selectors (Fig. 1).
11. Equilibrate the system by washing both flow-paths with a few mL each of DEPC-water, then stabilization buffer (*see* **Note 7**).
12. Obtain a stable jet at the desired combined flowrate by adjusting gas pressure and ethanol-sheath flow.
13. Commence mix-and-inject experiment with sample and ligand flowrates to achieve desired mixing-delay times.

3.2 Data Processing and Analysis

1. Data acquisition should be carried out with beam parameters appropriate for the given experiment, and guided by onsite data analysis (OnDA) using OnDA Monitor or equivalent [10].
2. Process data using Cheetah and CrystFEL for hit-finding and indexing, respectively [11–14]. Detector geometry should be optimized using CrystFEL geoptimiser [15]. Hit-finding and indexing parameters should be optimized to achieve the highest quality merged data for each individual time delay (*see* **Note 8**).
3. Once the final stream files are obtained for each data set, merge and scale the data sets using CrystFEL partialator with the custom-split option. This is necessary to ensure that the data being compared are processed and scaled in the same way prior to map calculation.
4. Calculate F_o–F_o DED maps for various sets of time-points, and investigate structural differences indicated by positive and negative DED peaks.
5. Where appropriate, model the structural differences at each time delay, and refine the models against their respective data sets.

4 Notes

1. In our experiments, we used 3D-printed micromixers with laminar stirring and DFFNs. In addition to helium gas, which is used for flow-focusing in standard GDVNs, DFFNs use an outer-sheath flow of ethanol to improve jet stability.
2. For micromixers, extra attention should be paid not only to the mixing speed but also to the spread of the delay time induced by the velocity profile of the liquid flow inside a capillary. Thus, micromixer designs that restrict the crystal flow to certain portions of the capillary radius are preferred.
3. The placement or integration of the micromixer into the flow scheme depends on the flowrates and desired mixing delay times. For shorter delay times (*e.g.*, milliseconds), the micromixer can be integrated into the DFFN. For longer delay times (*e.g.*, several seconds to minutes), the micromixer can be integrated upstream of the nozzle, either inside or before the nozzle rod, embedded in epoxy glue to improve resistance to high pressures.
4. For simplicity, and to avoid user error in channel selection, choose the same switchyard channel number for each reservoir type (*e.g.*, 1: sample/ligand, 2: stabilization buffer, 3: DEPC-water).

5. An inline filter (IDEX Health & Science, LLC, Oak Harbor, WA) can be installed in the sample line in addition to or in lieu of prefiltering to avoid clogging the nozzle. Alternatively, in the case of low hit-rates, the sample can be injected with no filtering, but at a great risk of clogging the nozzle.
6. To avoid introducing air into the system when connecting reservoirs to the flowpath selectors, pump through the corresponding channel at a very low flowrate.
7. This step should be repeated between samples and each time a reservoir is exchanged. This step can also be used in an attempt to purge a clogged line or nozzle and reestablish a stable jet.
8. The selection criteria to include more or fewer images in each stream file should be guided by figures of merit for merged data, and analysis of electron density maps. In cases of small individual data sets (*e.g.*, < 10,000 merged diffraction patterns), lower stringency conditions may be necessary to achieve higher multiplicity of data, but at the risk of noise. With high-multiplicity data sets, higher stringency conditions should be applied to minimize noise in the averaged reflection intensities. The number of patterns and signal-to-noise required for detection of differences in electron density maps is empirical, and will vary by experiment.

Acknowledgment

This research was supported by the Intramural Research Program of the NIH, National Cancer Institute, Center for Cancer Research (Y-X. Wang).

References

1. Kupitz C, Basu S, Grotjohann I et al (2014) Serial time-resolved crystallography of photosystem II using a femtosecond X-ray laser. Nature 513:261–265
2. Pande K, Hutchison CDM, Groenhof G et al (2016) Femtosecond structural dynamics drives the trans/cis isomerization in photoactive yellow protein. Science 352:725–729
3. Tenboer J, Basu S, Zatsepin N et al (2014) Time-resolved serial crystallography captures high-resolution intermediates of photoactive yellow protein. Science 346:1242–1246
4. Ramakrishnan S, Stagno JR, Conrad CE et al (2021) Synchronous RNA conformational changes trigger ordered phase transitions in crystals. Nat Commun 12:1762
5. Stagno JR, Bhandari YR, Conrad CE et al (2017) Real-time crystallographic studies of the adenine riboswitch using an X-ray free-electron laser. Febs J 284:3374–3380
6. Stagno JR, Liu Y, Bhandari YR et al (2017) Structures of riboswitch RNA reaction states by mix-and-inject XFEL serial crystallography. Nature 541:242–246
7. Schmidt M, Rajagopal S, Ren Z et al (2003) Application of singular value decomposition to the analysis of time-resolved macromolecular x-ray data. Biophys J 84:2112–2129
8. Zhao Y, Schmidt M (2009) New software for the singular value decomposition of time-resolved crystallographic data. J Appl Crystallogr 42:734–740

9. Knoska J, Adriano L, Awel S et al (2020) Ultra-compact 3D microfluidics for time-resolved structural biology. Nat Commun 11:657
10. Mariani V, Morgan A, Yoon CH et al (2016) OnDA: online data analysis and feedback for serial X-ray imaging. J Appl Crystallogr 49:1073–1080
11. Barty A, Kirian RA, Maia FR et al (2014) *Cheetah*: software for high-throughput reduction and analysis of serial femtosecond X-ray diffraction data. J Appl Crystallogr 47:1118–1131
12. White TA, Kirian RA, Martin AV, Aquila A, Nass K, Barty A, Chapman HN (2012) CrystFEL: a software suite for snapshot serial crystallography. J Appl Crystallogr 45:1–7
13. White TA (2019) Processing serial crystallography data with CrystFEL: a step-by-step guide. Acta Crystallogr D Struct Biol 75:219–233
14. White TA, Barty A, Stellato F et al (2013) Crystallographic data processing for free-electron laser sources. Acta Crystallogr D Biol Crystallogr 69:1231–1240
15. Yefanov O, Mariani V, Gati C et al (2015) Accurate determination of segmented X-ray detector geometry. Opt Express 23:28459–28470

Index

Jienyu Ding et al. (eds.), *RNA Structure and Dynamics*, Methods in Molecular Biology, vol. 2568, https://doi.org/10.1007/978-1-0716-2687-0,

MIX
Papier aus verantwortungsvollen Quellen
Paper from responsible sources
FSC® C105338

If you have any concerns about our products,
you can contact us on
ProductSafety@springernature.com

In case Publisher is established outside the EU,
the EU authorized representative is:
Springer Nature Customer Service Center GmbH
Europaplatz 3, 69115 Heidelberg, Germany

Printed by Libri Plureos GmbH
in Hamburg, Germany